CHIP-LEVEL MODELING WITH VHDL

JAMES R. ARMSTRONG

Virginia Polytechnic Institute and State University

PRENTICE HALL SERIES IN COMPUTER ENGINEERING
EDWARD J. McCLUSKEY, SERIES EDITOR

Prentice Hall, *Englewood Cliffs, New Jersey 07632*

Library of Congress Cataloging-in-Publication Data

ARMSTRONG, JAMES R. (date)
   Chip-level modeling with VHDL / James R. Armstrong.
   Includes bibliographies and index.

   ISBN 0-13-133190-6
   1. Integrated circuits--Computer simulation. I. Title.
TK7874.A75   1989   88-13995
621.381'73--dc19   CIP

Editorial/production supervision and
   interior design: Mary Rottino
Cover design: George Cornell
Manufacturing buyer: Mary Ann Gloriande

© 1989 by Prentice-Hall, Inc.
A Division of Simon & Schuster
Englewood Cliffs, New Jersey 07632

Printed in the United States of America
10 9 8 7 6 5 4 3 2

ISBN 0-13-133190-6

Prentice-Hall International (UK) Limited, *London*
Prentice-Hall of Australia Pty. Limited, *Sydney*
Prentice-Hall Canada Inc., *Toronto*
Prentice-Hall Hispanoamericana, S.A., *Mexico*
Prentice-Hall of India Private Limited, *New Delhi*
Prentice-Hall of Japan, Inc., *Tokyo*
Simon & Schuster Asia Pte. Ltd., *Singapore*
Editora Prentice-Hall do Brasil, Ltda., *Rio de Janeiro*

*This book is dedicated to my wife, Marie,
and my two children,
Michael and Anne Marie,
whose love makes all things seem possible.*

Prentice Hall Series in

 Computer Engineering

Edward J. McCluskey, Series Editor

ARMSTRONG	*Chip-Level Modeling with VHDL*
MCCLUSKEY	*Logic Design Principles*
RAO & FUJIWARA	*Error-Control Coding for Computer Systems*
WAKERLY	*Digital Design Principles and Practices*

CONTENTS

PREFACE

The purpose of this book is to introduce the concept of chip-level modeling, and thus allow designers to take a higher-level view of digital systems. The book is concerned with defining model structures that reproduce the I/O response of VLSI devices without modeling the details of internal structure.

The models in this book are expressed in VHDL, the VHSIC hardware description language. VHDL was chosen because it is an emerging standard for hardware description languages and, in the author's opinion, has the best set of constructs available for chip-level modeling. There are several different versions of VHDL available, which correspond to various stages of its development. The models in this book are written in the IEEE standard version of VHDL (1076B), which was released in 1987 and is the most up-to-date version of the language. However, at the time of the writing of this book, no simulator was available for the IEEE standard version. Thus the following procedure was used to validate the models in the book. All models were analyzed for correct syntax using the IEEE 1076 analyzer from CAD Language Systems Inc. In addition, versions of the models were also written in version 7.2 of the language and these models were analyzed and simulated using the version 7.2 analyzer and simulator developed by Intermetrics, Inc. Source code for the models is available from the author on IBM floppy diskette. The reader is encouraged to obtain a copy and analyze and simulate the models. The author would appreciate any feedback on the accuracy of the models that this process could give.

In writing this book, a particular viewpoint was taken. First, the author assumes a certain educational background: a knowledge of programming languages, such as Pascal or Fortran, and an understanding of basic computer organization and logic design. Also, while Chapter 1 provides some discussion of logic simulators, some background here would be very helpful. Thus, from a university instructional viewpoint, the book is intended for undergraduate seniors and graduate students in electrical engineering, computer engineering, and computer science. In fact, the book evolved from course notes on modeling developed for EE seniors and graduate students. Practicing engineers in the computer field should find their background adequate for the book.

Second, the book is written from the viewpoint of a hardware designer. This is particularly evident in the discussion of the application of VHDL to modeling. A more software-oriented person might use a given language construct as a unifying theme, and then cite various modeling situations as an example of the use of that construct. The viewpoint taken here was different: classes of model structures are presented, and the constructs applied to them, and in many cases, of course, the construct is reused.

Chapter 2 gives a summary of the VHDL language. However, it is difficult to cover the essential elements of a language in a single chapter. The reader is encouraged to consult the bibliography at the end of Chapter 2 for more detail.

The author wishes to acknowledge the help of the following people in the preparation of the manuscript:

1. Members of the EE5980 modeling class (fall 1986).   The notes for this class formed the basic outline for the book.
2. The reviewers.   Daniel Barclay, David Coelho, Walling Cyre, Dong Ha, Youm Hu, Bob Johnson, Scott Midkiff, Victor Nelson, Steve Piatz, Manos Roumeliotis, Larry Saunders, and Tuan Tran all contributed invaluable reviews of the rough draft of the book.
3. The model checkers.   A great deal of effort was expended in analyzing and simulating the models in the book. Dongil Han, Dave Burnette, Sandeep Shah, Khozema Khambati, and Greg Hulan provided invaluable aid in this process. They made many helpful suggestions which improved the models.
4. The figure drawers.   Nimish Modi and Dongil Han edited the figures on our computer graphics system.
5. My secretary.   Jenny Hedgepeth's skills with "PC script" made the manuscript preparation process a very efficient one.
6. My production editor.   Mary Rottino's patience and proficiency contributed to the smooth production of this book.

In addition to these people, I would like to thank the numerous members of the VHDL community who encouraged me throughout the project. Finally, I thank Paul Becker, my editor at Prentice Hall, for his enthusiastic support of the book.

1

INTRODUCTION

MOTIVATION: THE COMPLEXITY OF VLSI SYSTEM DESIGN

In this technological society in which we live, it is easy to get lost in detail, over-whelmed by the complexity we face. In mastering this complexity we must limit the amount of information analyzed at a given moment, that is, work at "the level of the forest instead of the trees." There is no more striking example of this than the problems faced by system designers in the era of very large scale integrated (VLSI) circuits. Since 1960, the number of gates on a chip has doubled approximately every two years. Today, individual chips contain hundreds of thousands of gates, and if one is attempting to assess the performance of a complicated system composed of these chips, it is of paramount importance that one maintain the proper level of view. Otherwise, the operation of these designs becomes very difficult to understand. The key to this problem is working at the proper level of abstraction— a level that provides the necessary information required at the moment, but does not overwhelm the designer with unnecessary detail.

THE ABSTRACTION HIERARCHY

In this section we present the abstraction hierarchy employed by digital designers. Abstraction can be expressed in two domains, which we now define as follows:

> *Structural domain*: a domain in which a component is described in terms of an interconnection of more primitive components

1

Behavioral domain: a domain in which a component is described by defining its input/output response by means of a procedure

In Figure 1.1 are shown the levels of abstraction in digital systems: processor memory switch (PMS), chip, register, gate, circuit, and silicon. In Figure 1.2 are shown examples of the representations at these levels. Referring again to Figure 1.1, note that the abstraction hierarchy has a pyramidal shape. The broadening of the pyramid as one moves to lower levels represents the increasing amount of detail that must be managed in representing a VLSI device at that level.

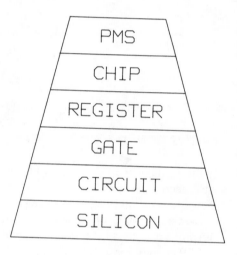

Figure 1.1   Abstraction levels in digital systems.

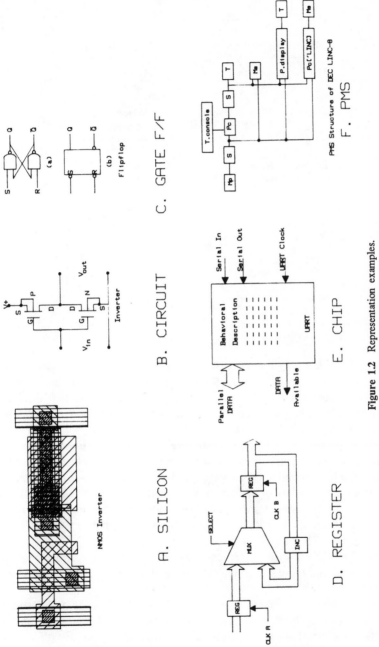

Figure 1.2 Representation examples.

3

Table 1.1   Nature of the Hierarchy

Level	Structural Primitives	Behavioral Representation
PMS	CPUs, memories buses	Performance specifications
Chip	Microprocessors, RAMs, ROMs, UARTs, parallel ports	I/O response, algorithms, micro-operations
Register	Registers, ALUs counters, MUXs	Truth tables, state tables, micro-operations
Gate	Gates, flip-flops	Boolean equations
Circuit	Transistors, R, L, and C	Differential equations
Silicon	Geometrical objects	None

In Table 1.1 is shown the nature of the hierarchy in terms of the structural and behavioral content of each level. At the lowest level, the silicon level, the basic primitives are geometric shapes that represent areas of diffusion, polysilicon, and metal on the silicon surface. The interconnection of these patterns models the fabrication process from the designer's point of view. The viewpoint here is structural only. At the next level up, the circuit level, the representation is that of an interconnection of traditional passive and active electrical circuit elements: resistors, capacitors, and bipolar and MOS transistors. The interconnection of components is used to model circuit behavior in terms of voltage and current. The behavioral content at this level can be expressed in terms of differential equations.

The third level up, the gate level, has traditionally been the major design level for digital design. The basic primitives are the AND, OR, and INVERT operators and various types of flip-flops. Interconnection of these primitives forms combinational and sequential logic circuits. Boolean equations are the behavioral content of this level.

The level above the gate level is the register level. Here the basic primitives are such things as registers, counters, multiplexers, and ALUs. These primitives are sometimes referred to as functional blocks; thus this level is also referred to as the functional level. Although the register-level primitives can be expressed in terms of an interconnection of gates, when working at this level one does not take this viewpoint. Register-level primitives are expressed in terms of truth tables and state tables; thus these two forms can be used to represent the behavioral content at this level. Register-level primitives can also be modeled behaviorally using register transfer languages.

The level above the register level is the chip level. As this level is the primary focus of this book, we define it in greater detail. At the chip level, the structural primitives are such things as microprocessors, memories, serial ports, parallel ports, and interrupt controllers. Although chip boundaries are typically the model boundaries, other situations are possible. For example, collections of chips which together form a single functional unit can be modeled as a single model, a good example of this being the modeling of a bit-slice microprocessor. Or alternatively, sections of a chip design could be modeled, perhaps during the design phase. The key aspect is that a large block of logic is to be represented in which long and frequently convergent data paths must be modeled from inputs to outputs.

As with the primitives at the lower levels, chip-level primitives are not constructed hierarchically from more basic primitives. Rather, they are a single model entity. Thus if a serial I/O port (UART) is to be modeled, the model is not composed by interconnecting simpler functional models of such things as registers and counters; the UART itself is the basic model. Models of this type are important to system manufacturers who buy a chip from another manufacturer but have no knowledge of its proprietary gate-level structure.

The behavioral content of a chip-level model is defined in terms of the I/O response of the device—the algorithm that the chip implements. The model is implemented as a sequence of micro-operations coded in a register transfer language. In Chapter 4 we define further the characteristics of chip-level models.

The top level in the structural hierarchy is the processor-memory-switch level. The primitive elements of this level are the processor, memory, and switch (bus). The behavioral content of this level is expressed in terms of performance specifications which give, for example, the MIP rating of a processor or the bandwidth in bits per second of a data path.

Note from Table 1.1 and the discussion above, that there is some overlap in the structural or behavioral characteristics of adjacent levels. For example, both the register and chip level can employ the micro-operation form of representation. However, the structural viewpoint is decidedly different, so we differentiate between them. The chip level and the PMS level have essentially the same primitive elements, but their behavioral content is completely different. The behavioral models at the chip level can compute detailed response in terms of integers or bits. The behavioral content of the PMS level is weak in that it serves chiefly as a bandwidth or stochastic model. In fact, PMS notation serves primarily as a classification scheme for computer architectures. In summary, a different level is required if either the behavioral or structural content is different.

The definition of an abstraction hierarchy for digital systems is a subject of continuing debate. One of the first hierarchies was that developed at Carnegie–Mellon University (CMU) by Bell and Newell. They specified levels consisting of circuit, gate, register transfer level, instruction set processor (ISP) level, high level language level, and PMS level. The circuit, gate, and register-level hierarchies defined here correspond to the circuit, gate, and register transfer levels of Bell and Newell. Both the chip level of our hierarchy and the ISP level of Bell and Newell have the basic property that the behavior content is specified by sequences of micro-operations. However, as will be demonstrated, chip-level models have the capability of modeling input/output timing accurately, whereas ISP models do not. ISP models were intended for the evaluation of instruction sets. Also, the term "instruction set processor" is specialized to the evaluation of computer architectures and is not appropriate for general hardware design. Finally, we do not have a high-level language level in our design hierarchy, again because our viewpoint is that of a hardware designer rather than a computer architect. This idea of defining the levels from the hardware designer's viewpoint is also shared by Walker and Thomas, also of CMU. In a paper delivered at the 1985 Design Automation Conference, they define a hierarchy with essentially the same levels as we have defined here. One difference is that they use the term "hardware module" instead of chip. To some people the term "chip" merely denotes a form of packaging that can be applied to a number of levels of hardware design. However, in the VLSI era, references to "chip-level phenomena" or "chip-level response" are frequently made by system designers, implying a chip-level model that involves a large amount of logic. Thus the term is a natural one from a designer's point of view and we will use it.

STRUCTURAL DESIGN DECOMPOSITION

The structural form of the design hierarchy implies a design decomposition process. This is because at any level we choose to model, the system model is composed by interconnection of the "primitives" defined for that level. An immediate question to be answered is: How are these primitives defined? The answer is that they are frequently defined in terms of primitives at the next lower level. Thus, as shown in Figure 1.3, a design can be represented as a tree, with the different levels of the tree corresponding to levels in the abstraction hierarchy. At the leaves of the tree, a change must take place, where the behavior of the lowest-level design components is specified.

As defined above, a behavioral model is a primitive model in which the operation of the model is specified by a procedure as opposed to redefining it in terms of other components. Since a behavioral model can exist at the circuit, gate, register, or chip level in a given design, different parts of the design can have behavior specified at different levels. In Figure 1.3a the design tree is "full" and thus all behavior is specified at the same level. In Figure 1.3b a design that has the form of a partial tree is shown, where behavior is specified at different levels. This situation is encountered because one frequently wants to evaluate the relationships between system components before they have been completely designed. Thus it is not required that all system components be specified at the gate level, for example, in order that the overall design be checked for errors. This checking is done by employing *multilevel simulation*, that is, employing a simulation in which the behavioral content of the component models is at different levels in the hierarchy. An important additional benefit of this approach is that it promotes simulation efficiency.

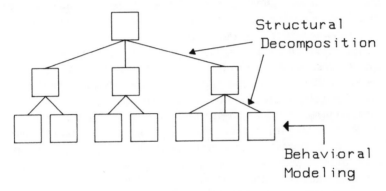

Figure 1.3(a) Full tree design.

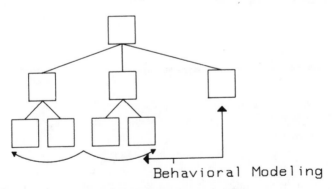

Figure 1.3(b) Partial tree design.

Two concepts related to the design tree are those of *top-down* and *bottom-up design*. Here the word "top" refers to the root of the tree, while "bottom" refers to the leaves. In top-down design, the design is begun with only the function of the root known to the designer. He or she then partitions the root into a set of lower-level primitives.

The design then proceeds to the lower level, where partitioning of the primitives at this level takes place. This process continues until the leaf nodes of the design are reached. An important point to make about top-down design is that the partitioning is optimized at each level according to some objective criterion. The partitioning is not constrained by "what's available."

The term "bottom-up" design is somewhat a misnomer in that the process of design still begins with the definition of the root, but in this case the partitioning is conditioned by what is available as primitives; that is, the designer must partition based on what parts will be available at the leaf nodes. These parts at the "bottom" will have been designed previously. Top-down design would appear to be the most ideal situation, but its disadvantage is that it produces components that are not "standard," thus increasing the cost of the design. Most real designs are therefore a combination of top-down and bottom-up techniques.

A final concept related to the hierarchy is that of a *design window*. By this we mean a range of levels over which the designer works in developing a design tree structure. The VLSI chip designer's window extends over the range of silicon, circuit, gate, register, and chip. The computer system designer, on the other hand, is currently concerned with a window consisting of the gate, register, chip, and PMS levels. It is this design window that is the main motivation for this book. For as the complexity of VLSI devices increases, it will become impractical to include the gate level in the design window, as individual chips will contain hundreds of thousands of gates. The register level, although certainly less complex than the gate level, may also offer unnecessary detail for those interested only in the input/output response of a VLSI chip. Thus the chip itself will become the primitive design element from the system designer's point of view.

SIMULATOR DESIGN CONSIDERATIONS

A prime motivation for employing chip-level models is to enhance simulation efficiency. Here we discuss the basic concept of a logic simulator and what considerations in its design affect efficiency. Figure 1.4 illustrates a typical logic simulator structure. Each device model, referred to here as a module, is represented by a procedure. Interconnection between devices is modeled by procedure calls. Events that occur within the simulator structure are kept in the simulator time queue. Time queue entries are represented as a three-tuple, the first integer representing the module number, the second the pin number, and the third the new logic value. The operation of the simulator proceeds as follows: The modules are numbered 0 through 3. Suppose that module 0 represents a primary input (PI). The initial event in the time queue (0, 6, 1) specifies that output pin 6 of module 0 should change to a logic 1 at time 0. When this occurs, a fan-out tracing program determines that this pin is tied to input pin 3 of module 1. If the value on the line is a new one, the procedure for module 1 will be invoked. Assume that it computes an output for pin 8 of module 1 and that due to the internal propagation delay that is modeled, that output is determined to change to a logic 0 100 nanoseconds (ns) after the present simulation time. This constitutes a new event that is scheduled by inserting it in the time queue. After all input changes at time 0 are processed, the simulation time is advanced to the next event time. The event processing continues until the event at time 100 is reached. Here the event (1, 8, 0) is processed: Module 1 pin 8 is set equal to a logic 0. If this is a signal change, the change is propagated to the inputs of modules 2 and 3 by the fan-out tracing program. The procedures for modules 2 and 3 are then invoked. Presumably, they compute new outputs and these are scheduled. Simulation continues until the event queue is empty or some externally controlled time limit is reached.

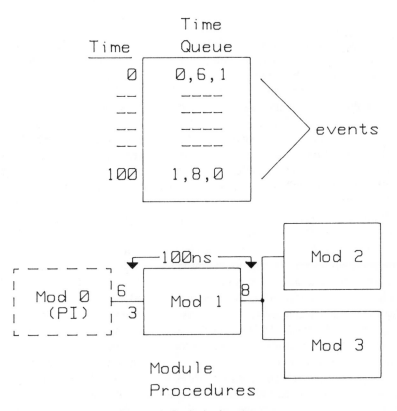

Figure 1.4 Logic simulator structure.

The example illustrates the first main point we wish to make about simulators and the related modeling process; that is, in creating models, we must have a scheduling mechanism. If the model is a chip-level model, the description will be expressed in a hardware description language. In this book we will be using the VHSIC hardware description language (VHDL). In VHDL one might write

$$X <= Y \text{ after } 100 \text{ ns};$$

and the effect would be to have X take on the value of Y at a simulation time 100 ns from the present time.

The second point to be made deals with simulation efficiency. When performing logic simulation, *simulation efficiency* (*E*) is defined as

$$E = \frac{\text{real logic time}}{\text{host CPU time}}$$

Real logic time is the actual time required to complete an activity sequence in a real logic circuit. *Host CPU time* is the time required to simulate the same activity sequence using a logic simulator running on a host CPU. To make the figures meaningful, the host CPU time must be normalized against the speed of the host. That is: Were the simulations run on a 1-MIP machine or a 10-MIP machine?

Another possible measure of efficiency is the number of events that a simulator can process per unit time. For example, gate-level simulators are frequently rated on the number of gate evaluations per host CPU second. For higher-level modeling, the definition of an event is defined in terms of a higher-level activity. For example, when modeling microprocessor systems, one might measure simulation efficiency in terms of the number of microprocessor clock pulses simulated per host CPU second. An even higher-level measure is used when modeling the execution of one CPU on another, that is, the number of (modeled) machine instructions simulated per host CPU second.

Simulation efficiency is determined by programming technique, the computer architecture employed, and the modeling level. By *programming technique* we mean whether the simulator is compiled or table driven. In the table-driven approach, the system interconnect is stored in tables and the models of the individual devices are called at run time as dictated by the interdevice signal flow. (The example in Figure 1.4 implied this approach.) In this case the simulation model is "interpreted" by the simulator at run time. In compiled simulators, the entire model is compiled into host machine code before simulation, and thus no run time interpretation is necessary. Compiled simulators are faster than table driven, but they suffer from portability problems. Table-driven simulators, on the other hand, are frequently written in a high-level language such as Fortran and Pascal and do not suffer from this difficulty. Some recent compiled simulators have solved the portability problem. For example, simulators for the VHSIC hardware description language compile the models into Ada or C and thus achieve portability.

The discussion of simulators above implicitly assumed the use of a uniprocessor host to run the simulator program. Recently, however, simulation engines have been developed which employ parallel array architectures to speed up gate-level simulations; for example, IBM's Yorktown Simulation Engine (YSE) and Zycad's simulation accelerator are capable of speeding up gate-level simulation by a factor of 1000.

Thus programming technique and host machine architecture can significantly affect simulation efficiency. However, the most fundamental property affecting simulation efficiency is the level in the representation hierarchy at which the simulation is performed. For example, circuit-level simulations employing SPICE, while yielding accurate results, are very inefficient; that is, only several hundred circuit nodes can realistically be simulated in a given circuit simulation run. Gate-level simulations can also be very inefficient. For example, a gate-level simulation of one 100 ns clock period of an Intel 8086 microprocessor can require 1 second of CPU time on a 1-MIP machine. This is a simulation efficiency of $10^{(-7)}$. In this case the simulation included detailed timing modeling. Unit-delay or zero-delay simulations at the gate level can achieve better efficiency than this.

Simulations at the chip level can achieve simulation efficiencies that are 100 to 10,000 times better than those at the gate level. Modelers should always choose the highest possible level in the hierarchy, where they can perform their simulations and still achieve the desired accuracy. Therefore, if one is interested only in the input/output response of chips in a system, modeling at the chip level is indicated.

SUMMARY AND OUTLINE OF SUCCEEDING CHAPTERS

In this chapter we have described the design hierarchy and some important characteristics of logic simulators that relate to the modeling process. We have proposed the use of chip-level models and given the following rationale for their use:

1. From the point of view of the system designer, gate and register-level models are too detailed for one wishing to maintain a proper perspective in analyzing system performance.
2. Lower-level simulations are too expensive for detailed simulation of total system response.
3. A system manufacturer may not have access to a gate-level model of a device. Chip-level models of the device will allow the manufacturer to assess system performance.

Thus the use of chip-level models to study the interaction between components in complicated systems is advocated. However, in order that the interaction be modeled accurately, the chip-level models must be accurate behavioral models that reproduce the chip's I/O behavior without resorting to a gate-level description of the device. It is the purpose of this book to explain how to do this modeling.

We conclude this chapter with a brief outline of the balance of the book. In Chapter 2 we provide an introduction to VHDL, the hardware description language used to express the models in the book. In Chapter 3 basic modeling of digital logic is discussed. In Chapter 4 we give a formal definition of a chip-level model and describe in detail this type of modeling structure. Principles of system modeling are presented in Chapter 5, which concludes with a comprehensive system example. In Chapter 6 we discuss other issues related to high level modeling of digital systems. The book is summarized in the Postscript.

BIBLIOGRAPHY

Armstrong, J. R., "Chip Level Modeling and Simulation," *Simulation,* pp. 141-148, October 1983.

Armstrong, J. R., "Chip Level Modeling of LSI Devices," *IEEE Transactions on Computer-Aided Design,* Vol. CAD-3. No. 4, pp. 288-297, October 1984.

Blank, T., "A Survey of Hardware Accelerators Used in Computer-Aided Design," *Design and Test of Computer.,* Vol. 1, No. 3, pp. 21-42, August 1984.

Miczo, A., *Digital Logic Testing and Simulation.* New York: Harper & Row, Publishers Inc., 1986.

Siewiorek, D., C. Bell, and A. Newell, *Computer Structures: Principles and Examples.* New York: McGraw-Hill Book Company, 1982.

Szygenda, S. A., and Thompson, E. W., "Digital Simulation in a Time-Based Table Driven Environment," *Computer,* Vol. 8, No. 3, pp. 24-48, March 1975.

Walker, R. A., and D. E. Thomas, "A Model of Design Representation and Systhesis," *Proceedings of the 22nd Design Automation Conference,* 1985, pp. 453-459.

2
BASIC FEATURES OF VHDL

THE ADVENT OF HARDWARE DESCRIPTION LANGUAGES

In the VLSI era a structured design process is required. In response to this need, considerable effort is being expended in the design of design aids. Hardware description languages are a specific example of this, as a great deal of effort is being expended in their development. Actually, the use of these languages is not new. Languages such as CDL, ISP, and AHPL have been used for the last 10 years. However, their primary application has been the verification of a design's architecture. They do not have the capability to model designs with a high degree of accuracy; that is, their timing model is not precise and/or their language constructs imply a certain hardware structure. Newer languages such as HHDL, ISP', and VHDL have more universal timing models and imply no particular hardware structure.

Hardware description languages have two main applications: documenting a design and modeling it. Good documentation of a design helps to ensure design accuracy. It also is important in ensuring design portability, that is, solving the "Tower of Babel" problem that can exist between manufacturers.

All useful hardware description languages are supported by a simulator. Thus the model inherent in an HDL description can be used to validate a design. Prototyping of complicated systems is extremely expensive, and the goal of those concerned with the development of hardware languages is to replace this prototyping process with validation through simulation. Other uses of HDL models are test generation and silicon compilation. These other uses are discussed in Chapter 6.

The use of hardware description languages in the design process implies a different approach to design than that used in the past. Until recently, digital designers were restricted to relying on a combination of word descriptions, block diagrams, timing diagrams, and logic schematics to describe their designs. Also, for a designer to be personally involved in the simulation process was considered to be an "anathema." "Those who can't design, simulate" was the attitude. The situation today is becoming markedly different. Designers have a great deal more training in software. They prepare their designs in a workstation environment, where the design is entered as an HDL source file, through schematic capture or some combination of the two approaches. Simulation is a tool frequently employed by the designers themselves to verify the correctness of the design as it evolves. Designers trained in the old school of design can have difficulty adjusting to this new approach. They may consider the use of a hardware description language unnecessary overhead. Hopefully, their attitudes will change as they see the results of the use of these languages—clearer and more error-free designs.

VHDL: THE VHSIC HARDWARE DESCRIPTION LANGUAGE

To illustrate the concepts of chip-level modeling, a specific hardware description language must be used. Until recently, there has not been a standard hardware description language, as there are standard programming languages such as Fortran and Pascal. However, beginning in 1983, the U.S. Department of Defense sponsored the development of the VHSIC hardware description language (VHDL). The original intent of the language was to serve as a means of communicating designs from one contractor to another in the Very High Speed Integrated Circuit (VHSIC) program. However, the design of the language has received input from many individuals in the computer industry and thus reflects a consensus of opinion as to what characteristics a hardware description language should have.

In August 1985, version 7.2 of the language was released, representing the completion of the first major stage of the language development. Version 7.2 was a complete language in that it comprehensively provided constructs for structural and behavioral modeling, as well as means to document designs. However, after the release of version 7.2 by the Department of Defense, the IEEE sponsored the further development of VHDL, the ultimate goal being the development of an improved, standard version of the language. The review process was completed by May 1987 and the language reference manual (LRM) released for industrial review. In June 1987 the eligible IEEE members voted to accept this version of VHDL as the standard version and in December 1987, it was officially so designated by the IEEE.

In this book we use VHDL as the working language to illustrate the concepts of chip-level modeling. This choice was made, not just because of the emerging popularity of VHDL, but because the author feels that the set of constructs available in VHDL are very effective for chip-level modeling. This conclusion was supported by Han in his master's thesis, "Hardware Description Languages for Chip Level Modeling," where he compared the effectiveness of various hardware description languages in performing chip-level modeling. He also concluded that VHDL was the most effective language for this purpose.

As stated above, there are two main versions of VHDL, version 7.2 and the IEEE standard version of the language. The VHDL used in this book is the IEEE standard version.

This chapter constitutes a brief introduction to VHDL. In subsequent chapters more advanced features of VHDL are introduced "on line" as the need arises. The author has found this to be an effective approach to the teaching of computer languages. For more on VHDL, the reader is encouraged to consult the bibliography at the end of the chapter.

DESIGN ENTITIES

In VHDL a given logic circuit is represented as a *design entity*. The logic circuit represented can be as complicated as a microprocessor or as simple as an AND gate. A design entity, in turn, consists of two different types of descriptions: the *interface description* and one or more *architectural bodies*. We illustrate this concept with an example. Consider the interface description for a circuit that counts the number of 1's in an input vector of length 3:

```
entity ONES_CNT is
  port(A: in BIT_VECTOR(0 to 2);C: out BIT_VECTOR(0 to 1));
  end ONES_CNT;
```

One can see that the interface description names the entity and describes its inputs and outputs. The description of interface signals includes the mode of the signal (i.e., in or out) and the type of the signal, in this case BIT_VECTOR(0 to 2) and BIT_VECTOR(0 to 1) (i.e., 3 and 2-bit vectors).

The interface description is also a place where documentation information about the nature of the entity can be recorded. For example, let us rewrite the interface description and include, in comment form, the truth table of the entity (see Figure 2.1).

```
entity ONES_CNT is
  port (A: in BIT_VECTOR(0 to 2); C: out BIT_VECTOR(0 to 1));

------  Truth Table:
---

---  ------------------------
--  | A2    A1    A0   |   C1   C0 |
    ------------------------
--  | 0     0     0    |   0    0  |
--  | 0     0     1    |   0    1  |
--  | 0     1     0    |   0    1  |
--  | 0     1     1    |   1    0  |
--  | 1     0     0    |   0    1  |
--  | 1     0     1    |   1    0  |
--  | 1     1     0    |   1    0  |
--  | 1     1     1    |   1    1  |
    ------------------------
end ONES_CNT;
```

Figure 2.1 Commented interface description.

Note that the truth table has been inserted as a comment in the description, that is, any line beginning with two dashes is interpreted as a comment. This is only one example of the type of information that can be added to an interface description. In later chapters we give other examples.

ARCHITECTURAL BODIES

The interface description basically defines only the inputs and outputs of the design entity. What is required, in addition, is a means of specifying the behavior of the entity. In VHDL this is done by specifying an architectural body. As will be demonstrated, this body can specify the behavior of the entity directly (i.e., be a primitive body) or the architectural body can be a structural decomposition of the body in terms of simpler components.

At the beginning of the design process, designers usually have an algorithm in mind that they would like to implement. Initially, however, they would like to check the accuracy of the algorithm without specifying the detailed implementation. Thus the first architectural body that would be implemented is one that is purely behavioral. Figure 2.2 shows such an architectural body for the ones counter.

```
architecture  PURE_BEHAVIOR of  ONES_CNT  is

begin

process(A)
  variable  NUM: INTEGER  range  0  to  3;
  begin
  NUM: = 0;
  for  I  in  0  to  2  loop
    if  A(I)  =  '1'  then
      NUM  :=  NUM  +  1;
    end  if;
  end  loop;
  case  NUM  is
    when  0  =>  C  <=  "00";
    when  1  =>  C  <=  "10";
    when  2  =>  C  <=  "01";
    when  3  =>  C  <=  "11";
  end  case;
end  process;

end  PURE_BEHAVIOR;
```

Figure 2.2   A purely behavioral description of the ones counter.

Note in the architectural body shown in Figure 2.2 that the loop scans the inputs A(0) through A(2) and increments the variable NUM whenever a particular bit is set. Based on the final value of NUM, a case statement then selects which bit pattern to transfer to the output.

This behavioral body describes the operation of the algorithm perfectly, but its correspondence to real hardware is weak. Specifically, one could ask such questions as: To what logic does the looping construct correspond? What is the delay through the circuit? This architectural body provides no answers to these questions. However, at the early stages of design, one is usually not concerned about these detailed considerations.

Continuing the example, suppose that the design of the ones counter advances to the logic design stage. Figures 2.3a and b show the Karnaugh maps for the two outputs, C1 and C0.

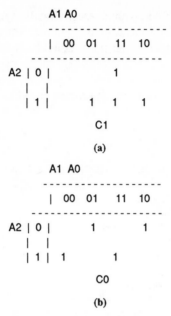

Figure 2.3   K-maps for the ones counter.

From these maps one can determine that the boolean equations for C1 and C0 are:

$$C1 = A1 \cdot A0 + A2 \cdot A0 + A2 \cdot A1$$

$$C0 = A2 \cdot \overline{A1} \cdot \overline{A0} + \overline{A2} \cdot \overline{A1} \cdot A0 + A2 \cdot A1 \cdot A0 + \overline{A2} \cdot A1 \cdot \overline{A0}$$

It should also be noted for later use that C1 is the majority function of three inputs A2, A1, and A0 [MAJ3(A2,A1,A0)] and C0 is the odd-parity function of three inputs [OPAR3(A2,A1,A0)].

At this point the designer could replace the purely behavioral body with the architectural body shown in Figure 2.4.

```
architecture TWO_LEVEL_MECH of ONES_CNT is

begin
  C(1) <= (A(1) and A(0)) or (A(2) and A(0))
          or (A(2) and A(1));

  C(0) <= (A(2) and not A(1) and not A(0))
          or (not A(2) and not A(1) and A(0))
          or (A(2) and A(1) and A(0))
          or (not A(2) and A(1) and not A(0));

end TWO_LEVEL_MECH;
```

Figure 2.4   Two-level mechanization of the ones counter.

In the statements implementing the "sum of products" equations for C(1) and C(0), the AND terms are enclosed in parentheses. This is because in VHDL, the AND and the OR operators are of equal precedence.

The two- level logic mechanization shown in Figure 2.4 implies a standard gate structure utilizing AND gates, OR gates, and inverters. However, since C1 can be computed by the MAJ3 function and C0 by the OPAR3 function an even simpler macro level architectural body would be

```
architecture MACRO of ONES_CNT is

begin

C(1) <= MAJ3(A);
C(0) <= OPAR3(A);

end MACRO;
```

This architectural body implies the existence of MAJ and OPAR gates at the hardware level. In terms of a VHDL description, it requires that the functions MAJ3 and OPAR3 must have been declared and defined previously. In a subsequent section we show how to do this.

The foregoing three architectural bodies for the ones counter are behavioral in that they specify the input/output response of the entity without exactly specifying the internal structure. We now give a structural architectural body for the ones counter. In the approach shown here, we will use a design hierarchy of the form shown in Figure 2.5; that is, the ones counter is first decomposed into MAJ3 and OPAR3 gates, which are in turn decomposed into AND and OR gates. Figure 2.6 shows the interface descriptions and architectural bodies for the MAJ3 gate, AND2 gates, and OR3 gates.

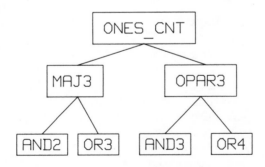

Figure 2.5 Structural design hierarchy for the ones counter.

We will explain architecture AND_OR of entity MAJ3 in detail. A logic diagram of its gate structure is shown in Figure 2.7. Referring again to Figure 2.6a, note that in the declaration section of architecture AND_OR, two components (AND2 and OR3) are declared. The port specification used is identical to the port specification used in the entity definitions for these components (see Figure 2.6b and c). Next, three signals (A1,A2,A3) are declared. The outputs of the three AND gates must be connected to the inputs of the OR gate, and the signal definitions allow us to represent these connections.

```
entity MAJ3 is
port (X: in BIT_VECTOR(0 to 2); Z: out BIT);
end MAJ3;

architecture AND_OR of MAJ3 is

component AND2
  port (I1,I2: in BIT; O: out BIT);
end component;
component OR3
  port (I1,I2,I3: in BIT; O: out BIT);
end component;
signal A1,A2,A3: BIT;

begin

G1: AND2
    port map (X(0) ,X(1) ,A1);
G2: AND2
    port map (X(0) ,X(2) ,A2);
G3: AND2
    port map (X(1) ,X(2) ,A3);
G4: OR3
    port map (A1,A2,A3,Z);

end AND_OR;
```

Figure 2.6(a)   MAJ3 description.

```
entity AND2 is
  port (I1,I2: in BIT; O: out BIT);
end AND2;

architecture BEHAVIOR of AND2 is
begin
  O <= I1 and I2;
end BEHAVIOR;
```

Figure 2.6(b)   AND2 description.

```
entity OR3 is
  port (I1,I2,I3: in BIT; O: out BIT);
end OR3;

architecture BEHAVIOR of OR3 is
begin
  O <= I1 or I2 or I3;
end BEHAVIOR;
```

Figure 2.6(c)   Description of OR3.

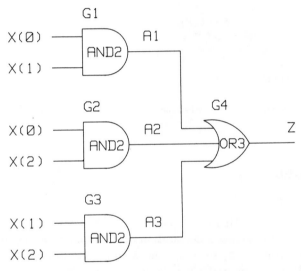

Figure 2.7   Majority function gate structure.

After the key word *begin*, four components are *instantiated*; that is, a specific instance of a general component entity is created. Note that each instantiation has a unique label associated with it (i.e., G1, G2, G3, or G4) as well as a port map. The port map creates an association between the inputs and outputs of the component declaration and the instantiated components. In this case the association is "by position." In Chapter Five we discuss another method.

To define the operation of entity MAJ3 completely, there must exist interface descriptions and architectural bodies for all components instantiated within the architectural body for MAJ3. Figure 2.6b and c gives these descriptions. The OPAR3 structure is similarly defined. We leave it as an exercise to the reader to do this. After the MAJ3 and OPAR3 components have been defined, a structural architectural description for the ones counter can be given and is shown in Figure 2.8.

```
architecture STRUCTURAL of ONES_CNT is

    component MAJ3
      port (X: in BIT_VECTOR(0 to 2); Z: out BIT);
    end component;
    component OPAR3
      port (X: in BIT_VECTOR(0 to 2); Z: out BIT);
    end component;

begin

    COMPONENT_1: MAJ3
      port map (A,C(1));
    COMPONENT_2: OPAR3
      port map (A,C(0));

end STRUCTURAL;
```

Figure 2.8   Structural architectural body for the ones counter.

BLOCK STATEMENTS

A basic element of a VHDL description is the *block*. A block is a bounded region of text that contains a declaration section and an executable section. Thus the architectural body itself is a block. However, within an architectural body, internal blocks can exist. Consider the example shown in Figure 2.9. Here blocks A and B are nested within the outer block of the architectural body. In general, any number of nesting levels is possible; for example, blocks A and B could also be decomposed into sub-blocks. There are two main reasons why this structure is employed. First, it supports a natural form of design decomposition, and second, a "guard" condition can be associated with a block. When a guard condition is TRUE, it enables certain types of statements inside the block. We shall see that guards are useful for modeling sequential logic.

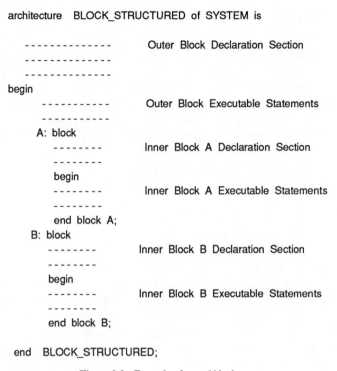

Figure 2.9 Example of nested blocks.

PROCESSES

Another major modeling element in VHDL is the *process*. Figure 2.2 gave an example of a process in which the behavioral description of the ONES_CNT circuit was given. Note that the process in this case began with the keyword *process(A)*. The vector A in this instance is said to make up the *sensitivity list* of the process; that is, whenever a signal in the sensitivity list changes, the process is activated and the statements within the process block are executed. A typical application of the process block is to implement algorithms at an abstract level, which of course is the case in Figure 2.2.

We discuss the implications of the process construct in Chapter 3 and discover that it represents the fundamental method by which concurrent activities in digital circuits are modeled.

DATA TYPES

Most programming languages support a variety of data types, and VHDL is no exception. However, because the language is used to represent hardware designs in a variety of ways, the data typing features are particularly important. For example, they give one the capability of representing a group of bus wires as (1) an array of bits, (2) an integer, or (3) a mnemonic code. Also, the language is *strongly typed*, which means that inadvertent mixing of types in an operation will be flagged as an error. The strong typing features are an important means for capturing the designer's intent.

We begin our discussion with the following definitions:

Type: a named set of values with a common characteristic

Subtype (of a Type):   a subset of the values of a type

So, for example, the set of all integers defines type INTEGER. While the set is infinite in size, from our point of view, the size of the set is constrained by host machine word length. The common characteristics shared by the elements of this set are their algebraic properties: (a) the operations defined on them (+, •), (b) an additive identity (0) and multiplicative identity (1), and (c) an additive inverse $[X + (- X) = 0]$. The natural numbers (i.e., all nonnegative integers) would define a subtype NATURAL of the type INTEGER. Note that the range of subtypes is constrained; for example, members of the subtype NATURAL cannot be negative integers. The subtype may or may not inherit all of the algebraic properties of the parent type. For example, for subtype NATURAL, + and • are defined, but there is no additive inverse.

Table 2.1 shows the predefined data types found in VHDL.

Table 2.1   VHDL Predefined Data Types

BOOLEAN	INTEGER	CHARACTER
BIT	POSITIVE	STRING
BIT VECTOR	NATURAL	
	REAL	

The left-hand column in the table gives the *logical* types. Type BOOLEAN consists of values TRUE and FALSE. All IF statements in the language must test objects or expressions of this type. Type BIT consists of the values '0' and '1'. Type BIT_VECTOR defines an array of bits, for example, BIT_VECTOR(0 to 7).

The middle column in Table 2.1 gives the *arithmetic* types and subtypes. Type INTEGER and subtype NATURAL were discussed above. Subtype

POSITIVE consists of all integers greater than zero. It may seem surprising to see the REAL type in a hardware description language. However, this type is very useful for high-level behavioral descriptions, for example, an analog-to-digital interface or a signal-processing algorithm.

The right-hand column in Table 2.1 gives the *character* types. Type CHARACTER essentially consists of the ASCII character set. Type STRING is an array of characters [e.g., STRING(0 to 9)]. Examples of these and other types are given below.

The discussion of data types so far has been from the user's point of view. The "VHDL Language Reference Manual" gives a useful classification scheme which is illustrated in Figure 2.10. Data types are classified as *scalar* (one-dimensional) or *composite* (multidimensional). The *enumeration* type is the most basic of the scalar types; that is, it merely enumerates the members of the type. For example, we will use a multivalued logic type MVL, which would be declared as follows:

type MVL is ('0','1','Z');

Note that the types BOOLEAN, BIT, and CHARACTER are enumeration types.

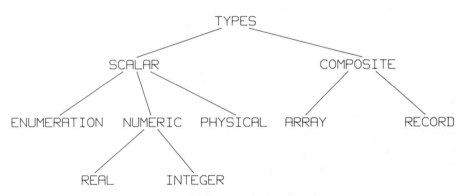

Figure 2.10   VHDL type classification scheme.

Numeric types are either INTEGER or REAL and require little further discussion except to give some examples:

```
type   INDEX is range 0 to 9;  - - - -  an integer type
type   VOLTAGE is range 0.0 to 10.0; - - - -  a real type
```

Each of the type declarations has a range. The base type (i.e., either INTEGER or REAL) is implied by the values in the range.

Because VHDL is a hardware description language, certain *physical* types are provided. For example, the type TIME is defined as follows:

```
type TIME is range 0 to 1E20
  units
    fs;
    ps  = 1000 fs;
    ns  = 1000 ps;
    us  = 1000 ns;
    ms  = 1000 us;
    s   = 1000 ms;
    min =   60  s;
    hr  =   60 min;
  end units;
```

Important features to note here are the base unit (femtoseconds) and range (1E20). This range and base unit allow the representation of 100,000 seconds (27.7 hours) as the maximum defined time. Both the base unit and the range are user selectable, but the range is constrained by the host machine word length. Other common physical types for hardware description languages would be voltage, current, power, weight, frequency, and so on.

Composite types are either *array* types or *record* types. Type BIT_VECTOR is an array type which is implicitly declared as follows:

```
type BIT_VECTOR is array( NATURAL range) of BIT;
```

BIT is the base type of the array elements. The expression "NATURAL range" is a general way of saying that the array length will be specified by a range of natural numbers. The user must specify this range when employing the type, for example, BIT_VECTOR(0 to 3) (ascending range) or BIT_VECTOR(7 downto 0) (descending range). Other array types with other base types are, of course, possible. For example, a descending array of integers could be defined by the following type declaration:

```
type INT_ARRAY is   array(99 downto 0) of INTEGER;
```

As the name implies, a record type is a composite type consisting of a number of fields. For example, type DATE could be defined as a record type as follows:

```
type DATE is
  record
    DAY :  INTEGER range 1 to 31;
    MONTH :  MONTH_NAME;
    YEAR :  INTEGER range 0 to 3000;
  end record;
```

Type MONTH_NAME would be an enumeration type consisting of the days of the month.

We said above that VHDL is a strongly typed language. Thus the source analyzer will flag as an error any attempt to mix different types. This includes, of course, user- defined types and subtypes. For example, a user could define the following:

```
subtype   ADDRESS is BIT_VECTOR(0 to 12);
subtype   DATA is BIT_VECTOR(0 to 15);
variable  PC: ADDRESS;
variable  ACC: DATA;
```

Then the statements

$$PC := ACC;   \quad \text{or} \quad   ACC := PC;$$

would be illegal since variables PC and ACC are of a different subtype.

Let us reinforce our understanding of types, subtypes, and type checking with one more example. Assume that the following type, subtype, and signal declarations have been made:

```
type X_INT is range 1 to 50;
type Y_INT is range 1 to 30;
subtype Z_INT is X_INT range 1 to 30;
signal  X: X_INT;
signal  Y: Y_INT;
signal  Z: Z_INT;
```

Then the question is: Which of the following signal assignment statements would be proper?

```
(2.1)      X <=  Y;
(2.2)      X <=  Z;
```

Statement 2.1 would be illegal since X and Y are signals of a different type. Statement 2.2 would be legal since Z_INT is a subtype of X_INT.

As has been implied above, to use types and subtypes, they must be declared. Entities, architectural bodies, blocks, processes, and subprograms all have declarative sections for this purpose. We give examples of this in the remainder of the book. We also see that they can be collected into "packages" for convenience.

In addition to the scalar and composite type classes discussed here, VHDL also allows file types and access types. However, as these types are not employed in the modeling discussed in this book, we do not treat them here.

OPERATORS

Corresponding to the data types available, VHDL has a complete set of operators. Table 2.2 provides a summary.

Table 2.2   VHDL Operators

Class	Class Members
Logical	not | and | or | nand | nor | xor
Relational	= | /= | < | <= | > | >=
Adding	+ | - | &
Signing	+ | - |
Multiplying	* | / | mod | rem
Miscellaneous	** | abs

The *logical operators* are defined for type BIT or type BOOLEAN, or for one-dimensional arrays in which each element is of type BIT or type BOOLEAN; thus they are defined for type BIT_VECTOR.

The *relational operators* all return type BOOLEAN as their result (i.e., the values TRUE or FALSE). All relational operators have a left and right argument (e.g., A /= B). The two equivalence checking operators (=, /=) can have any type as their left and right operands, provided that the types are the same. If the type is a composite type (e.g., an array), the testing is carried out on the elements of the composite. Two operands of composite type are defined to be equal if and only if each of their corresponding elements is equal.

The *ordering operators* (<, <=, >, >=) are defined for any scalar type; for example, given the enumeration type

```
type MVL is ('0','1','Z');
```

the ordering is from left to right, and the relation '1' 'Z' is true. The ordering operators are also defined for one-dimensional arrays in which each element is a discrete type. For example, suppose that we defined a three-element multivalued array type as follows:

```
type MVL is ('0','1','Z');
type THREE_BIT_MVL is array (0 to 2) of MVL;
```

Then the ordering operators may be applied to the three-element vectors. The comparison starts with the lowest index value and moves toward the highest index value. Thus for our example, ('0','1','1') <= ('0','Z','1') is TRUE.

The *arithmetic adding operators* (+, –) are defined for types INTEGER and REAL. They are not defined for type BIT_VECTOR and the user must provide subroutines for this purpose. The concatenation operator (&) has its normal function; that is, it combines two one-dimensional arrays to form a single one-dimensional array whose length is sum of the two operand arrays. Thus, for example, ('0','1','1') & ('0','Z','1') is equal to ('0','1','1','0','Z','1').

The *signing operators* (+, –) are unary operators used to precede single objects of type INTEGER or REAL with an arithmetic sign. Again, any complementation (ones or twos) of bit vectors must be done by user supplied subprograms. Both multiplication and division (*, /) are defined for types INTEGER and REAL. The remaining operators in the table [i.e., modulo (mod), remainder (rem), absolute value (abs), and exponentiation (**)] all have definitions normal to standard programming languages.

Operators and their operands appear, of course, in expressions [e.g., (A > B) and (C /= D)]. Many other examples follow.

CLASSES OF OBJECTS

In VHDL there are three classes of objects: constants, variables, and signals. An object is created when it is declared. As we discuss the three object classes, we give examples of these declarations.

Constants

A constant is an object whose value cannot be changed. Some examples of constant declarations are

```
constant TIEOFF   : MVL : = '1';
constant OVFL_MSG : STRING(1 to 20)
                    := "Accumulator Overflow";
constant INT_VECTOR : BIT_VECTOR(0 to 7) := "00001000";
constant COEFF: REAL := 4.217;
```

Note that each constant declaration gives the name of the constant, its type, and its value.

Variables

Variables are objects whose value can be changed. When a variable is created by declaration, a container for the object is created along with it. Variables are changed by executing a variable assignment statement (e.g., A := B + C;). Variable assignment statements have no time dimension associated with them (i.e., their effect is felt immediately). Thus variables have no direct hardware correspondence but are useful in algorithmic representations. Some examples of variable declarations are

```
variable  RUN : BOOLEAN := FALSE;
variable  COUNT : INTEGER := 0;
variable  ADDR : BIT_VECTOR(0 to 11);
```

Thus variable declarations specify the name, type, and optionally, an initialization value for the variable.

Variables used in a process block are considered to be *static*; that is, the value of the variable is maintained by the simulator until changed by a variable assignment statement. As an example, suppose that a process block was used to model the operation of a RAM memory:

```
RAM : process(ADDR,RW,CS)

    type MEMORY is array(0 to 1023) of BIT_VECTOR(0 to 7);
    variable MEM: MEMORY;

    begin
    - - - - - - - - - -      Read from and write to MEM.
    - - - - - - - - - -
    end process RAM;
```

Since variable MEM is static, the values stored in MEM will be preserved until changed by a variable assignment statement.

Signals

Signals are objects whose values may be changed and have a time dimension. Signal values are changed by signal assignment statements; for example,

```
A  <= B + C   after 50 ns;
D  <= E nor C;
```

Note that the signal assignment statement uses the <= symbol in order to differentiate it from the variable assignment statement. The "after 50 ns" in the first example means that A will potentially take on its new value 50 ns from the present simulation time. We say "potentially", because within the model there may be more than one signal assignment affecting A.

In the second example there is no after clause; this is equivalent to "after 0 ns." However, signal assignments are used to represent real circuit phenomena, so if there is no after clause or if the time argument of the after clause evaluates to zero, it is assumed that the signal takes on its new value *delta* time later. Delta is an arbitrarily small time greater than zero. We elaborate on implications of the timing of the signal assignment statement in Chapter 3.

Another area where signals differ from variables is that signals can have multiple containers called *drivers*, and the value of the signal is a function of all the drivers. Figure 2.11 shows an example of this. Shown are two processes, A and B, which each have signal assignments to a signal X. For each process that assigns to a signal, a driver is created to hold the result of that assignment. In this example the drivers are labeled Dax and Dbx. The value of the signal X is computed by a *bus resolution function* (F in this example). Bus resolution functions are user defined and are evaluated when one of the drivers of the signal receives a new value. The value of the signal is then updated to the value computed by the bus resolution function. In Chapter 5 we explain how to write bus resolution functions. The full implication of the definition of signals in VHDL will become clearer as we discuss modeling.

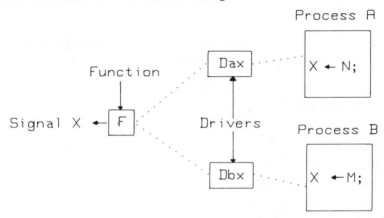

Figure 2.11   Multiple signal drivers and bus resolution function.

ATTRIBUTES

Attributes are values associated with a named entity in VHDL. Signal attributes are particularly important in modeling. Some examples:

1. S'EVENT is of type BOOLEAN. It is TRUE if an event has occurred on S during the current simulation cycle. It is useful for checking for changes in a signal [e.g., the statement if(S'EVENT) could check for changes in S].

2. S'STABLE(T) is of type BOOLEAN. It is TRUE if S has been stable for the last T time units. If T is zero, it is written as S'STABLE.

3. S'DELAYED(T) is the value of S T time units earlier. It has the same type as S.

These attributes are useful in detecting signal changes and detailed timing modeling, as illustrated in Chapter 4.

Another useful set of attributes are those associated with arrays. For example, suppose that an array variable was defined as follows:

variable A: BIT_VECTOR (0 to 15);

Then the following attributes would have the values indicated:

Attribute	Value
A'RANGE	0 to 15,
A'LEFT	0
A'RIGHT	15

Many other useful attributes are defined in the "VHDL Language Reference Manual." Users can also define their own attributes.

FUNCTIONS AND PROCEDURES

Functions can be declared in VHDL by specifying * :

1. The name of the function
2. The input parameters, if any
3. The type of the returned value
4. Any declarations required by the function itself
5. An algorithm for the computation of the returned value

In the architectural body MACRO for the entity ONES_CNT, the functions MAJ3(X,Y,Z) and OPAR3(X,Y,Z) were employed. To use these functions they must first have been declared as follows:

```
function MAJ3(X: BIT_VECTOR(0 to 2)) return BIT is
begin
  return (X(0) and X(1)) or (X(0) and X(2)) or (X(1) and X(2));
end MAJ3;
```

OPAR3 would be declared similarly. Note that the value returned is of type BIT. The value returned can be specified as an expression, which is the case above, or a sequence of statements can be used to compute the returned value. As stated above, local declarations can be made, but they are not required for this simple case.

---

* "VHDL User's Manual, Volume 1: Tutorial," IR-MD-065-1, August 1985, Intermetrics, Inc.

Procedures can also be written in VHDL. A procedure is declared by specifying *:

1. The name of the procedure
2. The input and output parameters, if any
3. Any declarations required by the procedure itself
4. An algorithm

Figure 2.12 shows the declaration of a procedure that does both ones and zeros counting on a 3-bit vector. Note that the parameter list specifies both input and output parameters. The parameter list is followed by a declaration section. The algorithm implemented by the procedure is the sequence of statements after the *begin* keyword.

```
procedure ONES_AND_ZEROS_CNT
(variable X : in BIT_VECTOR(0 to 2);
  variable N_ONES,N_ZEROS : out BIT_VECTOR(0 to 1)) is

  variable NUM1: INTEGER range 0 to 3 :=0;
  variable NUM0: INTEGER range 0 to 3 :=0;

begin
for I in 0 to 2 loop
if X(I) = '1' then
  NUM1 := NUM1 + 1;
else
  NUM0 := NUM0 + 1;
end if;
end loop;
case NUM1 is
  when 0 => N_ONES := "00";
  when 1 => N_ONES := "01";
  when 2 => N_ONES := "10";
  when 3 => N_ONES := "11";
end case;
case NUM0 is
  when 0 => N_ZEROS := "00";
  when 1 => N_ZEROS := "01";
  when 2 => N_ZEROS := "10";
  when 3 => N_ZEROS := "11";
end case;
end ONES_AND_ZEROS_CNT;
```

Figure 2.12 Procedure example.

* "VHDL User's Manual, Volume 1: Tutorial," IR-MD-065-1, August 1985, Intermetrics, Inc.

PACKAGES

It becomes tedious for the modeler to repeat declarations whenever she or he wants to use them. To alleviate this problem, VHDL uses a package mechanism for frequently used declarations. The package has a name associated with it. Declarations in the package may be made visible by referring to the package. First, however, the package must be declared. Figure 2.13 gives an example of such a declaration. Note that the package declaration for package HANDY declares subtypes and the interface for a function.

The code for the function is given in the package body. If a package contains no subprograms, a package body is not required.

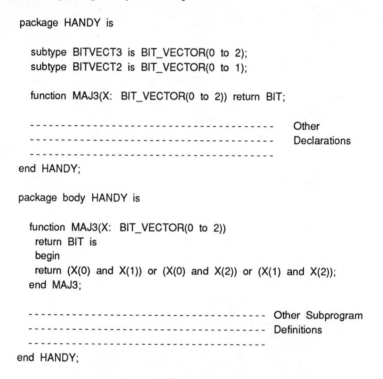

```
package HANDY is

    subtype BITVECT3 is BIT_VECTOR(0 to 2);
    subtype BITVECT2 is BIT_VECTOR(0 to 1);

    function MAJ3(X: BIT_VECTOR(0 to 2)) return BIT;

    ------------------------------------   Other
    ------------------------------------   Declarations
    ------------------------------------
end HANDY;

package body HANDY is

    function MAJ3(X: BIT_VECTOR(0 to 2))
    return BIT is
    begin
    return (X(0) and X(1)) or (X(0) and X(2)) or (X(1) and X(2));
    end MAJ3;

    ------------------------------------   Other Subprogram
    ------------------------------------   Definitions
    ------------------------------------
end HANDY;
```

Figure 2.13   Definition of a package.

Given the definition of package HANDY, suppose that one wanted to give an entity named LOGSYS access to the package. One could do this by placing a *use* clause before the interface description for entity LOGSYS:

```
use work.HANDY.all;
entity LOGSYS is
  port(X: in BITVECT3;Y: out BITVECT2);
  end LOGSYS;
```

All the declarations contained in HANDY would then be "visible" within the entity LOGSYS, including any of its architectural bodies.

Packages are a very useful language feature. Design groups can use standard packages that contain the type declarations and subprograms related to their system. As shown in Chapter 5, when we discuss system modeling, the amount of code in these declarations and subprograms is substantial. The package mechanism relieves the modeler from having to enter this code repeatedly. Also, if shared by different designers in a design group, it ensures consistency across the design.

The VHDL language defines a package STANDARD that can be used by all entities. Among other things, this package contains the definitions for types BIT, BIT_VECTOR, BOOLEAN, INTEGER, REAL, CHARACTER, STRING, and TIME, as well as subtypes POSITIVE and NATURAL.

CONTROL STATEMENTS

Table 2.3 shows the control statements in VHDL.

Table 2.3   Control Statements in VHDL

IF	LOOP	RETURN
CASE	NEXT	WAIT
	EXIT	

The full form of the IF statement is as follows:

```
if CONDITION_1 then
      sequence of statements
elsif CONDITION_2 then
      sequence of statements
else
      sequence of statements
end if;
```

Both CONDITION_1 and CONDITION_2 must be of type BOOLEAN. Any number of *elsif* clauses can be included. The *elsif* and *else* clauses are optional.

Two examples of case statements are:

```
Example 2.1                    Example 2.2

case   X(0 to 1) is            case N is
  when "00" => Z <= '0';         when 0 => Z <= '0';
  when "01" => Z <= '1';         when 1 => Z <= '1';
  when "10" => Z <= not Z;       when 2 => Z <= not Z;
  when "11" => Z <= Z;           when 3 => Z <= Z;
end case;                      end case;
```

The *case* statement performs decoding based on the value of a control expression and then executes a selected statement (or group of statements). In Example 2.1 the control expression is a bit vector; in Example 2.2 it is an integer. In general, the control expression can either be a discrete type, as the examples shown, or a one-dimensional array of characters.

Loop statements perform the function normal to all programming languages. Following are four examples:

Example 2.3   Example 2.4

```
for I in 0 to 3 loop                  SUM : = 0;
  A(I) := 2**I                        I := 1;
end loop;                    SUM_INT: while I <= N loop
                                        SUM := SUM + I;
                                        I := I + 1;
                                      end loop SUM_INT;
```

Example 2.5   Example 2.6

```
        SUM := 0;                     SUM := 1;
        I := 0;                       loop
SUM_INT: while I <= N loop            VAL(X);
        I := I+1;                     exit when X < 0;
        next SUM_INT when I = 3;      SUM := SUM + X;
        SUM := SUM + I;               end loop;
        end loop SUM_INT;
```

Example 2.3 illustrates the basic *for* loop; Example 2.4, the *while* loop and the use of a loop label. The loop in Example 2.4 computes the sum of the first N integers. In Example 2.5, the same sum is computed, except that 3 is not added in, as that iteration is skipped using the *next* statement. Example 2.6 illustrates a potentially infinite loop, which adds values, provided by a subroutine, to a sum. The loop will be exited when the value provided by the subroutine VAL(X) is negative.

The two remaining control statements are RETURN and WAIT. The RETURN statement is used in procedures and functions and has been discussed. As we shall see in Chapter 3, the WAIT statement is used to suspend a process for a period of time or until an event occurs.

SUMMARY

In this chapter we have introduced VHDL. Our purpose has been to give the reader a basic understanding of the language, particularly those features utilized in the models in later chapters. The presentation here is obviously not complete. Other language features will be illustrated as part of the modeling process. The models will also give complete examples, where all the necessary features of the language are pulled together. In understanding this chapter, the reader should have learned enough VHDL (1) to appreciate the usefulness of the language in the chip-level modeling process, and (2) to understand model structures. Those wishing to write their own VHDL models should consult the following bibliography.

BIBLIOGRAPHY

Han, D.,"Optimal Constructs for Chip Level Modeling," Masters thesis, Department of Electrical Engineering, Virginia Polytechnic Institute, August 1986.

IEEE Computer, February 1985, "Special Issue on Hardware Description Languages."

IEEE Design & Test of Computers," VHDL: The VHSIC Hardware Description Language," April 1986.

"VHDL Language Reference Manual, Draft Standard 1076/B," April 1987, CAD Language Systems, Inc.

"VHDL Language Reference Manual, Version 7.2," IR-MD-045-2, August 1985, Intermetrics, Inc.

"VHDL Tutorial for IEEE Standard 1076 VHDL," June 1987, CAD Language Systems, Inc.

"VHDL User's Manual, Volume I: Tutorial," IR-MD-065-1, August 1985, Intermetrics, Inc.

"VHDL User's Manual, Volume II: User's Reference Guide," IR-MD-065-1, August 1985, Intermetrics, Inc.

3

BASIC MODELING
TECHNIQUES

The major emphasis in this book is on chip-level models. However, all such models necessarily contain elements of basic combinational and sequential circuit models. In this chapter we explain how to develop these basic models.

MODELING CONCURRENCY

The modeling of logic circuits has a requirement that is not shared by many other modeling applications; that is, the model must include provision for concurrency of execution. This is because logic signals flow in parallel. Figure 3.1 illustrates this concept. Three logic blocks are shown. If one assumes that input set 1 and input set 2 are activated simultaneously, logic blocks 1 and 2 will be activated together. Logic block 3 will be activated as soon as either of the outputs from logic block 1 (Z1) or from logic block 2 (Z2) change. While signals are propagating through logic block 3, new input signal changes can be propagating their way through blocks 1 and 2. Thus signal flow can take place through all blocks simultaneously.

The hardware description language must have a mechanism for modeling this simultaneity. In VHDL, this requirement is handled by the process construct. Each process represents a block of logic, and all processes execute in parallel. (Of course, if the simulator is running on a single processor, they actually execute in turn, but the observed effect from a simulation point of view is that they execute in parallel.) In VHDL a process is activated when a signal in its sensitivity list changes. Figure 3.2 shows the corresponding VHDL process structures for the logic block structure in Figure 3.1. Note that the sensitivity list for the process contains, in general, the input signal set for the logic block. To be specific, logic block 1 might be represented by the process shown in Figure 3.3.

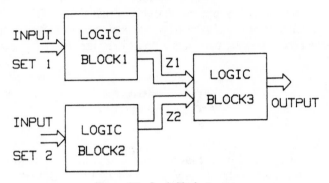

Figure 3.1   Logic block structure.

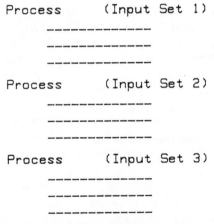

Figure 3.2   Three concurrent processes.

```
LOGIC_BLOCK1:  process(X1,X2,X3)
                 variable YINT: BIT;
               begin
                 YINT := X1 and X2;
                 Z1 <= YINT or X3 after 30ns;
                 end process LOGIC_BLOCK1;
```

Figure 3.3   Process example.

Note in the figure that the process body first contains a declaration section, where the variable YINT, which is "local" to the process, is declared. The executable section between the *begin* and *end* keywords consists of a variable assignment statement which computes an intermediate value (YINT) followed by a signal assignment statement which completes the evaluation of the function and incorporates the signal propagation delay across the block.

The idea of a process can also be incorporated into basic register transfer statements. Consider the following two register transfer statements:

A = X*Y;   ---- Statement S1
B = A+Z;   ---- Statement S2

The statements are printed in textual order, so our natural inclination is to say that they execute in sequence. If they do, notice that the value of A computed by S1 will be used as an input to statement S2; that is, the "new" value of A will be used as an argument for the execution of S2. This sequential execution is also referred to as "procedural" execution. However, these two statements can be interpreted another way. S1 and S2 can be considered processes. The sensitivity list for S1 contains X and Y, and the list for S2 contains A and Z. However, before we analyze the execution of S1 and S2 as processes, there is one other notion concerning a process that must be discussed. Since the process represents a physical system, it cannot execute in zero time. The example in Figure 3.3 illustrated how a propagation delay value can be assigned to a process. But in the absence of specifying a value we say that the process executes in "delta" time, a value of time that is infinitesimally small but greater than zero. Let us now consider the execution of S1 and S2, where they are regarded as processes. Assume that X and Z both change at time t_1. Then both S1 and S2 are activated. However, notice that in this case the value of A used by S2 is the value of A at t_1, which was determined by the values of X and Y at time t_1-delta. A new value of A will be available at t_1+delta and if this value reflects a change in A, S2 will be activated at that time.

In Figure 3.4a we illustrate this by means of an example. Assume that all the signals involved are of type integer. Recall that * and + denote integer multiplication and addition, respectively. At time t_1-delta the values of A and B are unknown since previous values of X, Y, and Z are unknown. At time t_1 the value of A is determined by the values of X and Y at t_1-delta; it is unaffected by the change in X that occurs at time t_1. The change in X at time t_1 has its effect on A at time t_1+delta and on B at time t_1+2*delta.

Figure 3.4b illustrates the case where the statements are interpreted to execute in the procedural sense (i.e., sequentially). Note that here the effect of changes is felt immediately (i.e., there is no delta delay).

	t_1-delta	t_1	t_1+delta	t_1+2*delta
X	1	4	5	5
Y	2	2	2	2
A	u	2	8	10
Z	0	3	2	2
B	u	u	5	10

Figure 3.4(a)   Process interpretation of S1 and S2.

	t_1-delta	t_1	t_1+delta
X	1	4	5
Y	2	2	2
A	2	8	10
Z	0	3	2
B	2	11	12

Figure 3.4(b)   Procedural interpretation of S1 and S2.

Returning again to the process interpretation of the statements, a basic question that must be answered is: What does delta represent?. An acceptable answer is that delta is one simulation cycle. Fundamentally, a simulation model consists of a set of processes. During the execution of a simulation cycle, all processes whose inputs have changed since the last cycle are evaluated. Signal outputs from those processes are scheduled to occur at later simulation times. Once all processes have been executed, the simulation cycle is complete. The next simulation cycle starts the next time a signal input to a process changes. That duration of time could be nanoseconds of simulation time, or it could merely mean that we have advanced to the next simulation cycle (i.e., delta time). The delta time duration results from signal assignments such as S1 and S2 which have no *after* clause.

From a logic circuit point of view, delta represents the delay through one stage of logic. Statements S1 and S2 represent a "data flow model" and the delta delay mechanism allows one to model the delay through the model without specifically assigning propagation delay values.

The issue of sequentiality versus concurrency is also addressed by the type of object employed in the register transfer statements. To illustrate this, let us refer again to statements S1 and S2 above. If we assume that the statements execute sequentially, then in the terminology of VHDL, A is a "variable" in that there is no time dimension associated with it (i.e., it takes on the value assigned to it immediately). If concurrent execution is assumed, A is a signal whose value can only change some finite amount of time (the minimum amount being delta) after an assignment is made to it.

Since one cannot, in general, infer from the text whether sequential or concurrent execution is implied, one must use either specialized notation or the semantics of a particular language to conclude which mode of execution is intended. In VHDL, specialized notation is employed. Concurrent execution is indicated through the use of the <= symbol, while procedural execution is implied by the := symbol. Thus for the example given above, if concurrent execution were implied, it would be written as

```
A <= X*Y;     ---- statement S1
B <= A+Z;     ---- statement S2
```

These two statements are termed signal assignment statements. Procedural execution would be implied by

```
A := X*Y;     ---- statement S1
B := A+Z;     ---- statement S2
```

where these two statements are termed variable assignment statements.

One more major question needs to be answered: Where can variable and signal assignments be used in VHDL descriptions? Variable assignments are restricted to use within processes (see Figures 2.2 and 3.3) and in subprograms (functions or procedures). Figure 2.12 is an example of subprogram usage. On the other hand, a signal assignment statement implies a process and thus can appear anywhere within the executable section of an architectural body. There is one situation that "muddies" the water a bit. Signal assignment statements can also appear within a process block. This is the case in Figures 2.2 and 3.3; let us use another example which better illustrates the implications of this:

```
process(R,S)
begin
    A <= X*Y;     --- statement S1
    B <= A+Z;     --- statement S2
end process;
```

The interpretation of this is as follows. The execution of statements S1 and S2 is not triggered by changes in X, Y, Z, or A. R and S comprise the sensitivity list for the process block. Thus when a change occurs on R or S, statements S1 and S2 effectively execute concurrently. The right-hand side of these statements will use the present values of X, Y, Z, and A to compute the values of A and B one delta time later. There is one restriction on the use of signal assignment statements within process blocks. There is only one driver defined per signal per process. Thus if two signal assignment statements within the same process assign to the same signal, a resolution function is not implied and the modeler must anticipate the possible effect of overwriting one signal value with another. However, as we shall see in subsequent modeling examples, this is generally not a problem because of the way the models are written.

One should realize that the nature of signal assignment statements, variable assignment statements, and process blocks give the modeler a number of choices for modeling. In the beginning of the design process he or she would use a process block that contains an algorithm expressed in variable assignment statements, the last statement in the block being a signal assignment statement that computes the output of the block. Figure 2.2 illustrates this. This is effective because no particular hardware structure is implied. The other approach is to use a set of signal assignment statements. Figure 2.4 shows this approach, which is sometimes referred to as a "data flow" implementation. This approach does imply a hardware structure and thus is more suitable for later phases of the design.

MODELING COMBINATIONAL LOGIC

In this section we present basic techniques for modeling combinational logic. Figure 3.5 illustrates the general form the circuit can take: An input vector X drives a logic circuit from which the output vector is F. For the circuit to be purely combinational it assumed that there is no feedback or charge storage mechanisms present which induce memory in the circuit. To illustrate the basic techniques, consider the simple case where we have a single output function $F = X1 \cdot X2 + X3$.

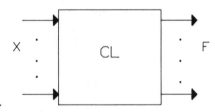

Figure 3.5   Basic combinational logic model.

A simple VHDL model of this function is

```
entity  CL  is
  port(X1,X2,X3:  in  BIT;
                  F:  out  BIT);
end  CL;

architecture  DATA_FLOW  of  CL  is
  signal T1:  BIT;
begin
  T1  <=  X1  and  X2;
  F  <=  T1  or  X3;
end  DATA_FLOW;
```

The two statements in the model are concurrent and therefore the model represents a data flow model that implies the gate circuit shown in Figure 3.6.   The delay through each gate is the arbitrarily small time value delta discussed above.   If one

wishes to model the time delay more exactly, the following two statement sequence
might be used:

```
T1 <= X1 and X2   after   DEL1;
F <= T1 or X3   after   DEL2;
```

In this case we are assigning specific propagation delays to each of the specific
gates. The values of DEL1 and DEL2 must be specified by use of a constant, expression, or generic declaration. We illustrate this in Chapter 4.

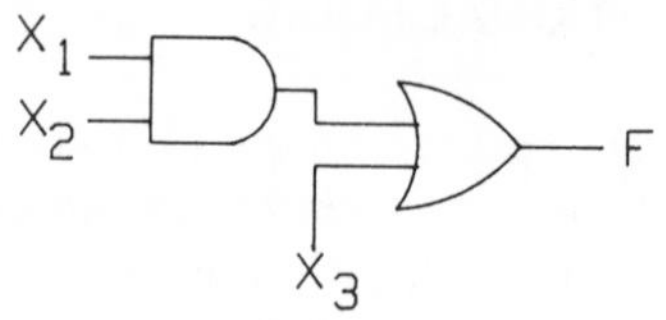

Figure 3.6 Gate level circuit for F.

An alternative representation for the contents of the block using the process
construct is as follows:

```
architecture PROCESS_IMPL of CL is
begin
  process (X1,X2,X3)
    variable T1: BIT;
  begin
    T1 := X1 and X2;
    F <= T1 or X3 after TOTAL_DEL;
  end process;
end PROCESS_IMPL;
```

In this case the process is activated whenever any signal within its sensitivity list
(X1, X2, X3) changes. Once the process is activated, the two statements execute
sequentially. T1 is a variable that is updated immediately (i.e., at the present simulation time without a delta delay). Next, the second statement executes and the value
of F is scheduled to change after time increment "TOTAL_DEL."

Other than the fact that this approach allows us to model the total delay versus the individual delays of the data flow model, there seems to be little benefit to
the process approach. However, consider the following alternative implementation
of just the process block in the same architectural body:

```
ROM:    process(X1,X2,X3)
           type FUNC_VALUES is array(0 to 7) of BIT;
           variable MEM: FUNC_VALUES:="01010111";
        begin
           F <= MEM(INTVAL(X1&X2&X3)) after TOTAL_DEL;
        end process ROM;
```

The implementation above implies a ROM implementation of the logic function. The ROM is modeled by a linear array of type FUNC_VALUES which defines a bit vector of length 8. The variable declaration specifies this type for the variable MEM and initializes MEM with the output values of the function. MEM is a linear array with an integer index. To access the array, the three inputs X1, X2, and X3 are concatenated together and then converted to type INTEGER, using the function INTVAL(VECT), which converts a bit vector to type INTEGER. How this function is implemented is described in Chapter 5. The ROM approach offers a great deal of flexibility, since one need only change the variable initialization and another combinational function is realized. PLA and PAL implementations of logic functions can use the same approach.

Another approach to combinational logic implies a multiplexer implementation:

```
MUX: process (X1,X2,X3)
     begin
       case X1&X2&X2 is
         when "000" => f <= '0' after TOTAL_DEL;
         when "001" => f <= '1' after TOTAL_DEL;
         when "010" => f <= '0' after TOTAL_DEL;
         when "011" => f <= '1' after TOTAL_DEL;
         when "100" => f <= '0' after TOTAL_DEL;
         when "101" => f <= '1' after TOTAL_DEL;
         when "110" => f <= '1' after TOTAL_DEL;
         when "111" => f <= '1' after TOTAL_DEL;
       end case;
     end process MUX;
```

In this implementation a pattern of ones and zeros (i. e., the values of f) are inputs to the multiplexer. The value of f is selected by the control inputs X1, X2, and X3.

VHDL provides a selected signal assignment statement which can implement the multiplexer somewhat more concisely:

```
with X1&X2&X3 select
  f <= '0' after TOTAL_DEL  when "000",
       '1' after TOTAL_DEL  when "001",
       '0' after TOTAL_DEL  when "010",
       '1' after TOTAL_DEL  when "011",
       '0' after TOTAL_DEL  when "100",
       '1' after TOTAL_DEL  when "101",
       '1' after TOTAL_DEL  when "110",
       '1' after TOTAL_DEL  when "111";
```

The selected assignment statement above just replaces the process statement in the architectural body.

MODELING SEQUENTIAL LOGIC

In the preceding section we defined a combinational logic circuit to be one that contains no feedback or memory storage. Circuits in which these elements are present are termed *sequential circuits*. Sequential circuits can be represented by the Huffman model, which is shown in Figure 3.7. The basic elements of the model are a block of combinational logic and a set of feedback loops. The combinational logic receives as inputs the external input vector X and the present value of the circuit state YDEL. Its outputs are the circuit output Z and the next value of the state variables Y. The delay can either be clocked (synchronous) or pure propagation delay (asynchronous). We consider first an example of the synchronous case. Figure 3.8a shows the block diagram of an UP counter module. When the control input, CON, equals 1, the counter counts up at 2 MHZ; when CON equals 0, the state of the counter does not change.

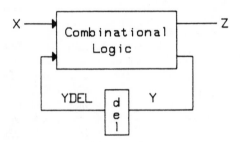

Figure 3.7   Huffman model of sequential circuit. (© 1988 IEEE).

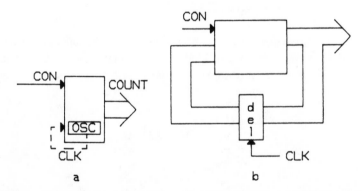

Figure 3.8   Huffman model of up counter. (a: Up counter; b: Huffman model) (© 1988 IEEE).

Figure 3.8b shows the block diagram for the Huffman model of the counter. The combinational logic section of the counter consists of an incrementer which is enabled by the CON input. The feedback delay would be implemented with clocked flip-flops. The incrementing function can be implemented with the combinational

logic modeling described above. The full model of the counter, which was
developed by Dongal Han, is as follows:

```
entity COUNTER is
generic(CLOCK_WIDTH,COUNT_DEL:  TIME);
port(CON: in BIT;
        Z: out BIT_VECTOR(0 to 3));
end COUNTER;

use work.INC_PAC.all;
architecture HUFFMAN of COUNTER is
 signal CLK: BIT:= '0';
begin
 process(CON,CLK)
  variable COUNT: BIT_VECTOR(0 to 3):= "0000";
 begin
  if not CON'STABLE then
   if CON = '1' then
    CLK <= transport '1' after CLOCK_WIDTH;
   else
    CLK <= transport '0';
   end if;
  end if;
  if (CLK ='1') and (not CLK'STABLE) then
     COUNT := INC(COUNT);
    CLK <= transport '0' after CLOCK_WIDTH/2;
    CLK <= transport '1' after CLOCK_WIDTH;
   end if;
   Z <= COUNT after COUNT_DEL;
  end process;
end HUFFMAN;
```

The model represents the sequential nature of the counter in two areas. First, the
state of the counter is stored in the variable COUNT. Since all processes in VHDL
are "alive" for the duration of the simulation run, the value stored in the variable
counter will be maintained. Second, the clocking action of the built-in oscillator in
the counter is modeled by the delayed assignment of '0' and '1' to the signal CLK,
and the fact that a change in CLK will activate the process that generated it. This
models the delayed feedback mechanism of the Huffman model. Note that the oscillator is started by the rise of CON and turned off when CON falls. Finally, the
assignments to the signal clock contain the keyword *transport*. This is done so that
one assignment statement does not override another. This will become clearer in
Chapter 4, when we discuss inertial and transport delay.

Asynchronous circuits can be implemented in a similar fashion, with the process construct allowing flexibility in specifying the level at which the circuit is modeled. Consider the following flow table for a certain asynchronous sequential circuit:

| | Q(t + 1) | |
Q(t)	x = 0	x = 1
y1 y2	Y1 Y2	Y1 Y2
0 0	0 0	0 1
0 1	1 1	0 1
1 0	0 0	1 0
1 1	1 1	1 0

Initially, the modeler may not want to specify the physical implementation of the circuit. Following is a VHDL model that stores the flow table in an array:

```
entity  ASYNC is
  generic (PROP_DEL:  TIME);
  port(X : in BIT;
       Y1,Y2: inout BIT);
end ASYNC;

use work.INTPAC.all;
architecture ROM of ASYNC is

begin
  process(X,Y1,Y2)
    type SQ_ARRAY is array(0 to 7, 0 to 1) of BIT;
    variable MEM: SQ_ARRAY:=(('0','0'),('1','1'),('0','0'),
                             ('1','1'),('0','1'),('0','1'),
                             ('1','0'),('1','0'));
  begin
    Y1 <= MEM(INTVAL(X&Y1&Y2),0) after PROP_DEL;
    Y2 <= MEM(INTVAL(X&Y1&Y2),1) after PROP_DEL;
  end process;
end ROM;
```

When the array is accessed, the first dimension index is computed by concatenating X, Y1, and Y2 and then converting this three-position bit vector to type INTEGER. Note that the outputs Y1 and Y2 are of mode inout in the interface description. This allows them to be accessed by the single statement processes in the architectural body, as well as being driven by those processes.

As the design is carried forward, boolean equations are developed for the state variables Y1 and Y2:

$$Y1 = \overline{x} \cdot y2 + x \cdot \underline{y1}$$
$$Y2 = \overline{x} \cdot y2 + x \cdot \overline{y1}$$

An alternative approach, then, to modeling the circuit at this phase in the design is expressed in the following architectural body:

```
architecture GATE_LEVEL of ASYNC is
begin
   Y1 <= (not X and Y2) or (X and Y1) after  DEL1;
   Y2 <= (not X and Y2) or (X and not Y1) after DEL2;
end GATE_LEVEL;
```

The examples above show that the feedback mechanism is modeled identically for synchronous and asynchronous circuits. Only the semantics of the feedback variable change; for synchronous circuits the feedback variable represents a clock period, while for an asynchronous circuit it represents propagation delay.

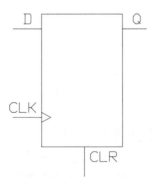

Figure 3.9 D latch.

CLOCKED FLIP-FLOP BEHAVIOR

Another basic feature of sequential circuits that must be modeled is that of clocked flip-flop behavior. In Figure 3.9 is shown the block diagram of a D latch. When the clock (CLK) is high, the output Q follows the input D; when CLK falls, the present value of Q is "latched in." When CLK is low, the stored value of Q is unchanged. The latch also contains an asynchronous clear (CLR). When CLR = 1, the latch is reset. The asynchronous clear overrides the clock. A VHDL block modeling the behavior of the latch is

```
D_LATCH: block (CLK = '1'or CLR = '1')
        begin
        Q <= guarded  '0' when CLR = '1' else
                      D;
        end block D_LATCH;
```

In VHDL the block header "block (CLK = '1' or CLR = '1')" implies that "CLK = '1' or CLR = '1'" is a "guard" expression. The guard is implicitly defined to be of type BOOLEAN. Any guarded statement will be executed when (1) the guard is TRUE and a signal on the right-hand side of the guarded statement changes, or (2) the guard changes from FALSE to TRUE. When the guard is false, the signal on the left-hand side retains its old value. For the example shown above, assuming that CLR is a "0," when CLK is a "1," the output Q will follow the input D. When CLK falls, the last value stored in Q is retained. Thus the latch behavior is implemented. Note that the clear function will be activated whenever CLR = 1; thus the clear function is asynchronous.

In contrast to the latch, which is level sensitive to the clock, many clocked flip-flops are edge-triggered. Below is a VHDL description of an edge-triggered version of a D flip-flop:

```
EDGE_TRIGGERED_D: block ((CLK = '1' and not CLK'STABLE)
                         or CLR = '1')
        begin
        Q <= guarded '0' when CLR = '1' else
        D;
        end block EDGE_TRIGGERED_D;
```

Notice that the guard expression is now

(CLK = '1' and not CLK'STABLE) or CLR = '1'.

'STABLE is an attribute of the signal CLK. As we discussed in Chapter 2, for a signal X, X'STABLE(T) will be TRUE if and only if X has been stable for the last T time units. X'STABLE(0), which is written more briefly as X'STABLE, will be FALSE if and only if X has just changed. Thus assuming that CLR = '0', the guard expression will be TRUE when the clock has just changed from 0 to 1.

FINITE STATE MACHINE REPRESENTATION

Designers frequently wish to use a finite state machine representation for a sequential circuit. As an example of this representation we consider a model of a finite state machine prepared by Michael Blau, a student in the chip level modeling class at Virginia Tech. Figure 3.10 shows the state diagram for a three-state Mealy machine which outputs a 1 whenever the last two inputs are equal. The VHDL description for the machine is given in Figure 3.11. The Mealy machine representation implies a synchronous circuit, so guarded blocks are used. For the outer block (B1), the guard is TRUE when the CLOCK makes a 0-to-1 transition. The guard expression for the inner blocks (S0, S1, S2) is an AND of the present state and the outer block guard. An enumeration type STATE is used to represent the three states of the machine. The state movement of the Mealy machine is represented by the guarded conditional signal assignments statements within the blocks S0, S1, and S2. The state of the machine is held in the signal STATE_REG, which is driven by the three signal assignment statements. Thus a bus function is required. This function, STATE_RES_FUNC, is declared at the beginning of block B1. Note that the signal STATE_REG is designated as being a register. This means that the driver of a guarded assignment statement affecting the signal will be ignored by the bus resolution function when the guard for that assignment statement is false. Since, in this case, all drivers but one will be ignored, the bus resolution function selects the active driver and uses its value as the output of the function. In Chapter 5 we discuss this concept in detail.

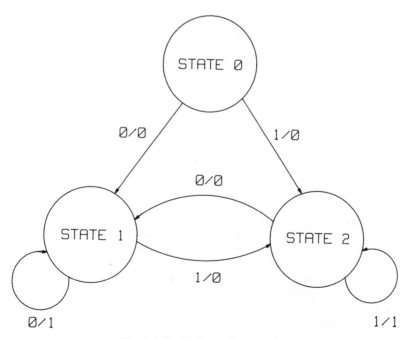

Figure 3.10 Mealy machine example.

```
entity MEALY_MACHINE is
 port (CLOCK, INPUT: in BIT;
         OUTPUT: inout BIT);
end MEALY_MACHINE;

architecture STATE_SEQUENCE of MEALY_MACHINE   is
begin
 B1: block(CLOCK='1' and not CLOCK'STABLE)

   type STATE is (STATE0,STATE1,STATE2);
   type RES_TYPE is array (INTEGER range < >) of STATE;

   function STATE_RES_FUNC (signal INPUT: RES_TYPE) return STATE is
    variable RESOLVED_VALUE: STATE;
   begin
    for I in INPUT'RANGE loop
     RESOLVED_VALUE := INPUT(I);
    end loop;
    return RESOLVED_VALUE;
   end STATE_RES_FUNC;

   signal STATE_REG: STATE_RES_FUNC STATE register := STATE0;

  begin
   S0: block((STATE_REG = STATE0) and guard)
   begin
    STATE_REG <= guarded  STATE1   when INPUT='0'
               else STATE2;
   end block S0;
   S1: block((STATE_REG = STATE1) and guard)
   begin
    STATE_REG <= guarded STATE1 when INPUT='0'
               else STATE2;
   end block S1;
   S2: block((STATE_REG = STATE2) and guard)
    begin
     STATE_REG <= guarded STATE1 when INPUT='0'
               else STATE2;
    end block S2;
 -- output
    OUTPUT <= '0' when STATE_REG=STATE0 else
              '1' when STATE_REG=STATE1 and INPUT='0' else
              '0' when STATE_REG=STATE1 and INPUT='1' else
              '0' when STATE_REG=STATE2 and INPUT='0' else
              '1' when STATE_REG=STATE2 and INPUT='1' else
              OUTPUT;
  end block B1;
end STATE_SEQUENCE;
```

Figure 3.11   VHDL description of a mealy machine.

USE OF THE WAIT CONSTRUCT

In digital systems a logical process will frequently pause in its execution while waiting for a time period to elapse or an event to occur. Once the time period has elapsed or the awaited event has occurred, execution of the process resumes. This situation is illustrated as follows:


```
process
_____        -----Begin  process  execution

_____

_____
wait  for  time  T
       or
wait  until  event  E
_____        ------Resume  process  execution

_____

_____
end  process           ------End  process  execution
```

Although it may not be apparent, the use of the wait mechanism implies sequential logic behavior. Being in a wait state until either a time-out or an event occurs implies sequential state storage and sequencing logic. If a time-out is used for resuming the process, counting, another sequential activity, is implied. Finally, since the process resumes where it left off after the wait period (as opposed to restarting), it is required that the process state present before the wait be preserved. This state includes a pointer to the next activity to be carried out.

The form of the wait construct in VHDL is

```
WAIT    on    sensitivity    until    condition    for    time  ;
              list                     clause              out
```

The statement suspends the process until a signal in the sensitivity list changes, at which time the condition clause is evaluated. The condition clause is an expression of type BOOLEAN. If it is TRUE, the process resumes. The time-out clause sets the maximum wait time after which the process will resume. As an example:

```
WAIT on X,Y until (Z = 0) for 100 ns;
```

would suspend a process until either X or Y changed, then the expression $Z = 0$ is evaluated, and if TRUE, the process will resume. Regardless of the conditions above, the process will resume after a delay of 100 ns.

A WAIT statement can contain one or more of the condition statements; for example,

(Example 3.1)   WAIT on X,Y;
(Example 3.2)   WAIT until (Z = '0');
(Example 3.3)   WAIT for 100 ns;

are all valid wait statements. In Example 3.1 the condition clause is assumed to be TRUE and the process will resume when X or Y changes. In Example 3.2, since there is no sensitivity list, the variable in the condition clause is assumed to make up the sensitivity list (i.e., the process will resume when Z changes from a 1 to a 0). In Example 3.3, the process will resume after 100 ns, irrespective of any other conditions.

There is one restriction on the use of the WAIT statement in VHDL. In VHDL there are two forms of the process statement:

```
process(X,Y,Z)              process
------------                ----------

------------                ----------
end process;                end process;

Form   A                    Form  B
```

Form A is a process that becomes active when a signal in its sensitivity list changes. Form B has no sensitivity list and implies a process that is always active. Form A is actually equivalent to the following:

```
process
---------

------------
WAIT on X,Y,Z
end process;
```

Thus when simulation begins, the process executes once and then waits at the end of the process for signal changes that will cause it to resume. The restriction on the modeler is that WAIT statements cannot appear in form A processes, since they would always be executed before the implicit WAIT statement at the end of the process, destroying the intent of the form A process. However, since form B processes can have as many WAIT statements as one wishes, this does not really limit modeling capability.

Two major applications of the wait construct in modeling are the modeling of component interaction and oscillator behavior. We give an example of component interaction first:

```
INTERFACE: process
           begin
           wait on A;

           ----------
           ----------   Set Up Interface Control
           ----------   Signals
           ----------

           wait until (B = '1') for 100ns;

           ----------
           ----------   Act on response
           ----------

           end process INTERFACE;
```

In the example process above, INTERFACE is always active but waits on the signal A to initiate its first action, which is to set up interface control signals. It then waits for B to change from a 0 to 1 or for the time-out period before acting on the response.

Next we illustrate the modeling of an oscillator using the wait construct:

```
process
begin
for I in 1 to NUM_CLK loop
  CLK <= transport '1' after CLK_PERIOD/2;
  CLK <= transport '0' after CLK_PERIOD;
  wait for CLK_PERIOD;
  end loop;
end process;
```

Note that the process above continues to generate clocks until NUM_CLK is reached. However, most of the process lifetime is spent in the wait state.

The wait construct was originally developed for modeling at the computer architecture level. As such, its hardware correspondence is not closely defined. If the intended use of the description is not merely to model good behavior, but to do fault modeling or perhaps compilation of the description into "silicon," the modeler might be better served by using a lower-level description.

SUMMARY

In this chapter we have described basic techniques for modeling digital logic. An important concept in this modeling process is concurrent execution, which models the effect of parallel signal flow. We presented two concepts of delay: delta delay and timing delay. We discussed various approaches to modeling of combinational and sequential circuits at varying levels of abstractions. We showed how the wait construct could be used to model sequential circuit phenomena. The techniques described in this chapter will be used to model elements of the chip-level structures described in the remainder of the book.

BIBLIOGRAPHY

Hill, Frederick J. and Gerald R. Peterson, *Introduction to Switching Theory & Logical Design*, 3rd Ed., New York: John Wiley & Sons, Inc., 1981.

"VHDL Language Reference Manual, Draft Standard 1076/B," April 1987, CAD Language Systems, Inc.

"VHDL Tutorial for IEEE Standard 1076 VHDL," June 1987, CAD Language Systems, Inc.

"VHDL User's Manual, Volume 1: Tutorial," August 1985, Intermetrics Inc.

4

CHIP–LEVEL
MODELING

In this chapter we define the characteristics of chip-level models, explain their internal structure, and show how this structure is implemented in VHDL. The modeling of timing is important in chip-level models and we will see how timing accuracy is incorporated into these models. The concepts of chip-level modeling will be illustrated by means of three examples.

DEFINITION AND GENERAL CHARACTERISTICS OF A
CHIP-LEVEL MODEL

Definition: A *chip-level model* is a behavioral model of a block of logic in which signal path delays are accurately modeled without resorting to lower-level descriptions.

This definition implies the following characteristics:

1. Chip-level models model input/output timing accurately. We shall see that the term "accurate timing" can have a number of interpretations.

2. Chip-level models are implemented as sequences of micro-operations coded in a hardware description language. As such, they are not expressed as a decomposition of simpler structural primitives. It is assumed that the chip itself is a basic primitive (i.e., represents a leaf on the design tree).

3. The block of logic can be of arbitrary size, but the most useful application of the modeling technique will be the modeling of LSI and VLSI circuits which contain thousands of gates. Model boundaries can be the interface defined by the pin-out of a real chip (e.g., a microprocessor, UART, or programmable

parallel port). The model could also represent a collection of chips that perform a system function (e.g., a bit-slice processor). In this second case one might model the collection of chips as a whole because one was interested in the I/O response of the overall function, not the detailed timing between the individual chips in the collection. A third possibility is that a portion of what ultimately would become a VLSI chip would be modeled, perhaps in the early stages of the chip design.

4. Chip-level models can be constructed without a gate- or register-level description. The modeler can use word descriptions, block diagrams, timing specifications and diagrams, and state tables and state diagrams. The accuracy of the model is directly related to the accuracy of this information. The fact that chip-level models can be constructed without lower-level descriptions has (at least) two major applications. First, early in the design, one can check the performance of proposed system components in relation to other components, even though a lower-level implementation does not exist. Timing information cannot be completely accurate at this point, but knowledge of the target technology (e.g., NMOS, CMOS, or ECL) can allow some reasonable inference of timing delays across blocks. A second major application of chip-level models is after the chip has been manufactured. System manufacturers will use this device as a component and will wish to model their systems. However, in many cases the gate-level model is proprietary to the manufacturer. Even if it is available, it could be prohibitively expensive to simulate with this low-level model. Companies that market simulators currently offer behavioral models of register level components. A few chip-level models of such VLSI components as microprocessors are also available.

5. A chip-level model can be expressed as a set of specifications that define a chip-level view of an LSI or VLSI device. Figure 4.1 illustrates this. Input specifications describe setup time, hold time, and minimum pulse-width restrictions on inputs to the chip. Internal specifications describe the micro-operations of the device, its internal memory requirements, and major internal path signal delays. The output specifications determine the time at which outputs switch.

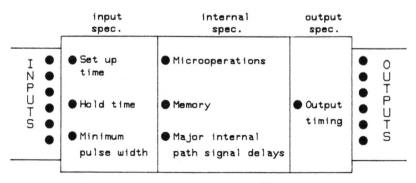

Figure 4.1   Chip–level view of a VLSI device. (© 1988 IEEE).

6. For the timing to be accurate, a fine-grained timing model is required. As pointed out in Chapter 2, some early hardware description languages assumed coarse timing models. This is not adequate for our definition of a chip-level model. As an example, consider Figure 4.2, which shows typical timing for a digital system. Note that although output signal changes and input sampling are initiated by internal clock changes, they actually occur between clock transitions because of internal signal propagation delay. Thus a timing model which assumes that all output signal changes and input sampling occur in coincidence with clock transitions is not accurate enough. More recent hardware description languages, such as ISP', HHDL, and VHDL, have accurate timing constructs to handle this situation.

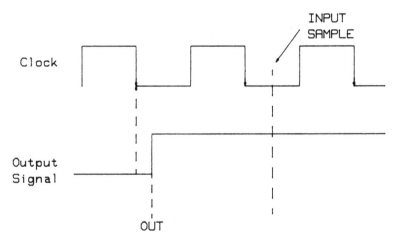

Figure 4.2   Digital system timing. (© 1988 IEEE).

We have described in this section the general characteristics of chip-level models. In the next section we will see that these characteristics imply a certain model structure.

CHIP-LEVEL MODEL STRUCTURES

We have defined chip-level models as being behavioral models. However, we shall see that they have a structure. Not structure in the sense of physical decomposition (for then they would not be behavioral) but structure in the sense that the model can be represented as a directed graph. Other important aspects of the model are that it represents (1) delay, (2) minimum energy, and (3) functional partitioning.

MODELING DELAY

When one performs chip-level modeling, there are three generic delay structures that are quickly encountered [*] (see Figure 4.3):

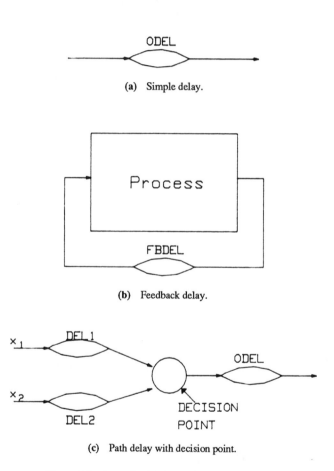

(a) Simple delay.

(b) Feedback delay.

(c) Path delay with decision point.

Figure 4.3 Generalized delay models. (© 1988 IEEE).

---

[*] Armstrong, J. R., "Chip Level Modeling With HDLs," IEEE Design and Test of Computers, February 1988, pp. 8-18.

1. *Simple delay.* Simple propagation delay (Figure 4.3a) from an input to an output is the most straightforward delay structure. This is modeled with the delayed assignment of one variable or signal to another signal. Delayed assignments are implemented by scheduling a queue event to occur. All HDLs that do not assume ideally synchronous timing can implement simple delay. As described in Chapter 3, in VHDL one can write

$$X <= Y \text{ after 35 ns;}$$

And X will follow Y after 35 ns.

2. *Feedback delay.* Feedback delay (Figure 4.3b) was also discussed in Chapter 3 as an essential element of sequential circuit modeling. The combination of the delayed assignment statement and the process construct of VHDL, with an output connected back as an input to the process's sensitivity list were the necessary requirements for implementing this structure.

3. *Path delay with decision point.* A more complex delay model is the *path delay with decision point model* shown in Figure 4.3c. Here inputs X1 and X2 activate separate delay paths (DEL1 and DEL2) which converge at a decision point. At this point a logical decision is made whether to propagate a new value forward through the output delay path ODEL. This structure and variations on it are very commonly encountered in chip-level models. For example, consider the case of the tristate buffered register in Figure 4.4. Path DEL1 in this case is the data path, activated by the strobe signal, which propagates the input data to the latch outputs and the tristate buffers' inputs. The DEL2 path is the path from the selection puts to the tristate buffers. ODEL is the path delay through the output buffers. For a change in either the strobe or the select inputs, a new value of the vector (signal) must be propagated to the tristate buffers and then a logical decision made as to whether a new value should be propagated to the output.

In a chip-level model, the decision point is internal to the model; thus there must exist a mechanism internal to the model for waking up the decision point. In modeling with VHDL, a "process" mechanism can be used to implement the *path delay with decision point model*. In the VHDL approach, the total chip behavior is represented by an architecture, with the processes within the architecture representing computational nodes. Signals representing propagation delay paths are passed between the processes.

Figure 4.5 shows the path delay with decision point model, in which computational nodes have been designated and processes assigned to them. Process 1 monitors inputs X1 and X2, propagating the delayed values (X1DEL and X2DEL) of any input changes. Process 2 consists of the decision point and associated output delay. Its inputs are X1DEL and X2DEL. When process 2 is invoked, the decision point may, depending on its function, compute a new value of Z' and propagate this value through the output delay (ODEL) to form the output signal Z.

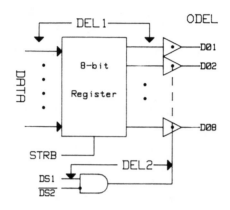

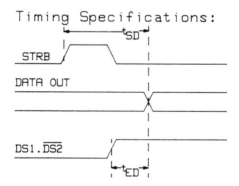

Figure 4.4   Delays through tri-state buffered register. (© 1988 IEEE).

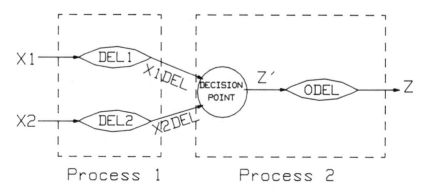

Figure 4.5   Process assignment for delay with decision point. (© 1988 IEEE).

```
entity  DELAY_WITH_DECISION  is
  generic (DEL1,DEL2,ODEL:  TIME) ;
  port (X1,X2: in BIT; Z: out BIT) ;
end DELAY_WITH_DECISION ;

architecture PROCESS_IMPL of DELAY_WITH_DECISION is

signal X1DEL, X2DEL: BIT;

begin

INPUT:
process(X1,X2)
  begin
    if  (X1'EVENT) then
      X1DEL <= X1 after DEL1;
    endif;
    if  (X2'EVENT)
      X2DEL <= X2 after DEL2;
    endif;
  end process INPUT;

DECISION:
process (X1DEL,X2DEL)
  variable Z_PRIME: BIT;
  begin
  ------            decision making
  ------            performed here.
  ------
    Z <= Z_PRIME after ODEL;   -- output propagation
  end process DECISION;

end PROCESS_IMPL;
```

Figure 4.6   VHDL process implementation of delay with decision.

Figure 4.6 shows the VHDL description of the path delay with decision point model. First, the entity describes the model interface, followed by an architectural body giving a behavioral description of the model. In the body, the internal signals X1DEL and X2DEL are first declared. Next, after the *begin* keyword, the behavior of the body is defined in terms of two processes: INPUT and DECISION. For each process the sensitivity list indicates which signal changes will activate the process. In process INPUT the code checks for changes in X1 and X2, and if either of those changes occur, it propagates the new values to X1DEL and X2DEL. X1DEL and X2DEL are in the sensitivity list for process DECISION, so when one of them changes, this process will be activated and a decision made as to whether a new value should be propagated to the output.

PROCESS MODEL GRAPHS

The structures that we have discussed above are generic in the sense that they relate to basic delay structures that one would wish to model. However, from a graph-
theoretic point of view, they are just specific examples of a general structure present
in all chip-level models. In Figure 4.7 is shown a graph representation of a typical
chip-level model. The graph represents a partitioning of the function of the model
into subfunctions, with each node representing a subfunction. Each subfunction can
be implemented by a VHDL process. The arcs denote signal passage between the
process nodes. Each arc is labeled with a designator of the form S(DEL_S), where
S is the signal name and DEL_S is the delay in transmitting a signal from one process
to another. We shall refer to this graph representation as the *process model graph*
of a chip-level model. It is a very useful representation for chip-level models, as it
allows us to model complex signal flow within models while using a high-level representation. Examples of the use of this graph are given in the modeling examples
that follow shortly.

FUNCTIONALLY PARTITIONED MODELS

The process model graph defined above has a graph form controlled by the signal
paths in the chip and the corresponding delays. However, models can also be partitioned functionally. This functional partitioning can be applied to the nodes in the
process model graph. Figure 4.8 illustrates this concept. Node C has been decomposed into three functions: F1, F2, and F3. When modeling in VHDL, one frequently assigns a process to each function. Thus the model for node C might be

```
process (X, Y)        ------ function F1
   -------------

   -------------
end process


process (X, Z)        ------ function F2
   -------------

   -------------
end process


process (Y, Z)        ------ function F3
   -------------

   -------------
end process
```

Note from Figure 4.8 that node C has three inputs X, Y, and Z, and that in the
example above each of the processes is activated by a subset of these inputs. In the
concluding example of this chapter we illustrate the concept of functional partitioning of an node.

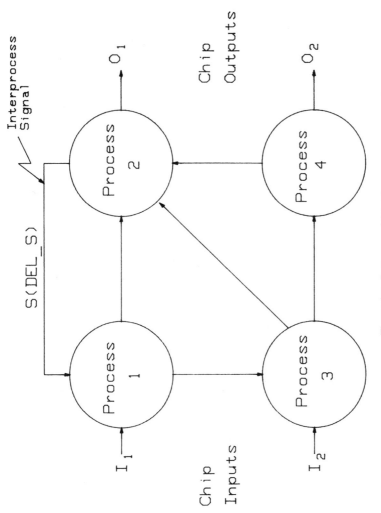

Figure 4.7 Process model graph.

Interprocess
Signal

S(DEL_S)

Process
1

Process
2

Process
3

Process
4

I_1

I_2

O_1

O_2

Chip
Inputs

Chip
Outputs

65

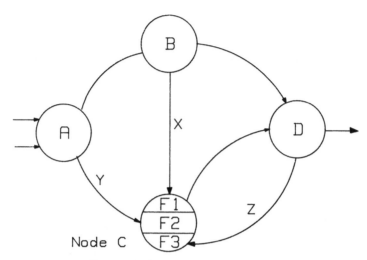

Figure 4.8 Functional partitioning of a node.

In some models, functional partitioning is the only form of partitioning employed. In Chapter 5 we give an example of this.

TIMING MODELING

The definition of a chip-level model requires an accurate representation of input/output timing. We now will begin to define what is meant by the term "accurate."

In VHDL there are two types of delay in signal assignment statements, *inertial delay* and *transport delay*. Shown below are two examples of this:

```
X  <=  Y    after 3 ns;  ----inertial delay
X  <=  transport Y after 3 ns; ----transport delay
```

The first assignment statement implies inertial delay; that is, the signal propagation will only take place if and only if an input persists at a given level for a specified amount of time. In VHDL this specified amount of time is the delay given in the *after* clause.

Thus in the example above, changes in Y will affect X only if they stay at the new level for 3 ns or more. In the second case, where transport delay is specified, all changes on Y will propagate to X regardless of how long the changes stay at the new level.

The inertial delay mechanism filters out inputs that change too rapidly. What is modeled is a circuit effect. A logic signal is represented by a node voltage at the circuit level. Because of capacitance, node voltages cannot change instantaneously. They require a certain amount of energy to persist for a given amount of time for the voltage to change enough to cause the circuit driven by the voltage to switch. In VHDL, inertial delay is implied by a signal assignment statement unless the key word "transport" is used. The use of inertial delay is quite natural when modeling real hardware. Transport delay would more likely be employed at higher levels of abstraction.

TIMING ASSERTIONS

The built-in inertial delay feature of the signal assignment statement will filter out a pulse on a signal shorter than the minimum-duration specification. However, there is no indication to the modeler that this condition has occurred, other than the fact that the signal has not changed. It is useful to be able to print an error message when the minimum pulse-width specification is violated. Also, in many cases, timing constraints are more complex in that they involve the relationships between signals. For example, consider the clocked register shown in Figure 4.4. It is a common requirement that the data input be stable for a duration of time prior to the clock transition that strobes data into the register. This requirement is known as the *setup time* specification on the data relative to the clock. A similar requirement states that the data should remain stable a minimum amount of time after the clock makes its transition (i.e., a *hold-time* specification). VHDL allows us to handle this situation with assertions that have the following form:

assert   Boolean Expression

report   Error Message;

When the assertion is executed, it checks the boolean expression. A TRUE value indicates a normal condition. A FALSE value indicates an error condition and the error message is printed. Assertions can be used to check practically anything, but in this application we are using them to check the timing of signals. Thus for timing assertions, the components of the boolean expressions are signal attributes. Two particularly useful signal attributes for this purpose are X'STABLE(T) and X'DELAYED(T). As explained in Chapter 2, X'STABLE(T) is TRUE if and only if signal X has been stable for T time units. X'DELAYED(T) is the value of X T time units earlier.

We illustrate the use of assertions in timing modeling with the following example. Figure 4.9 shows typical input specifications for the register in Figure 4.4. The specification says that (1) DATA should be stable for S nanoseconds before STRB rises (setup time), (2) DATA should be stable for H nanoseconds after STRB rises (hold time), and (3) STRB should have a minimum duration at the one level of W nanoseconds (minimum pulse width). The following assertion checks the setup-time specification:

```
assert not (not STRB'STABLE and (STRB = '1') and not DATA'STABLE(S))
report  "Setup Time Failure";
```

Thus if the strobe has just made a positive transition, and the data input has not been stable for the previous S nanoseconds, a setup-time failure will be reported.

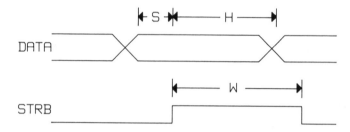

Figure 4.9 Input timing specification.

Using DeMorgan's theorem, one could convert the assert statement to the simpler form

```
assert STRB'STABLE or (STRB = '0') or DATA'STABLE(S)
```

Again using DeMorgan's theorem, one could also write the simplified form for the hold-time test:

```
assert  STRB'DELAYED(H)'STABLE or (STRB'DELAYED(H) = '0')
        or DATA'STABLE(H)
report  "Hold Time Failure";
```

Similarly, for the minimum pulse-width check, the assertion is

```
assert  STRB'STABLE or (STRB = '1') or STRB'DELAYED'STABLE(W)
report "Minimum pulse width failure";
```

Assertions can be inserted in the interface description or the architectural body of a design entity. If placed in an architectural body, the assertions are specific to a particular implementation of the design entity. If placed in the interface description, they can be used to check the timing of signals in and out of any architectural body of the design entity. Below we give examples of both approaches.

MODELING EXAMPLES

Example 1

As our first example of a chip-level model, consider the 8-bit register shown in Figure 4.4. Its timing specification is also shown in the figure. The register is loaded on the rise of the strobe (STRB) and assuming that the output buffers are enabled, the output of the register will change Tsd nanoseconds later. The enable condition for the register buffer is the AND of the DS1 and $\overline{DS2}$ inputs. Any change in the enable condition will cause the outputs to change Ted nanoseconds later.

We use this example to illustrate three levels of timing modeling. The first level is a model that reproduces the behavioral response of the register but does not consider timing at all:

```
process (STRB,DS1,NDS2)
begin
  if STRB = '1' and not STRB'stable then
    REG <= DI;
    if DS1 = '1' and NDS2 = '0' then
      DO <= DI; else
      DO <= "11111111";
    end if;
  elsif not DS1'stable or not NDS2'stable then
    if DS1 = '1' and NDS2 = '0' then
      DO <= REG; else
      DO <= "11111111";
    end if;
  end if;
end process;
```

The only delay exhibited by this process is delta delay. We shall refer to this type of timing as *delta delay timing*. Models with delta delay timing are useful for initial system architectural verification.

The next level of timing modeling is what we shall refer to as *simple I/O timing*:

```
process (STRB,DS1,NDS2)
begin
  if STRB = '1' and not STRB'stable then
    REG <= DI after STRB_DEL;
    if DS1 = '1' and NDS2 = '0' then
      DO <= DI after STRB_DEL + ODEL; else
      DO <= "11111111" after STRB_DEL + ODEL;
    end if;
  elsif not DS1'stable or not NDS2'stable then
    if DS1 = '1' and NDS2 ='0' then
      DO <= REG after EN_DEL + ODEL; else
      DO <= "11111111" after EN_DEL + ODEL;
    end if;
  end if;
end process;
```

The model assumes that Tsd = STRB_DEL + ODEL and Ted = EN_DEL + ODEL. STRB_DEL is the delay from the rise of the strobe to the arrival of the data at the register output. EN_DEL is the delay from the enable inputs to the output buffers; ODEL is the delay through the output buffers.

This second model was developed merely by adding delays to the assignment statements in the delta delay model. It seemingly reproduces the I/O timing response, but it has one deficiency: if the strobe (STRB) and either DS1 or NDS2 change simultaneously, the model is inaccurate. We thus designate the timing as being "simple I/O timing" because internal data paths are modeled in a simplistic (and in this case inaccurate) fashion.

The model using simple I/O timing does not model data paths accurately because it uses a single process. We will now present a model that corrects this deficiency and employs *data path timing*. This level of timing modeling fully meets the requirement for accurate timing modeling in chip-level models. Figure 4.10 shows a process model graph for the register. Process A receives the strobe and data inputs and outputs a delayed value of the data (REG). Process B similarly produces a delayed value of the enable signal (ENBLD). Process C receives REG and ENBLD and produces the output DO. The delays for the three signal paths (STRB_DEL, EN_DEL, ODEL) are also indicated on the graph. The full VHDL description for the model employing data path timing is shown in Figure 4.11. In the interface description, the three delays in the model have been specified as "generic." This allows the model to be general. When the model is instantiated in an interconnected system description, the value of these generic parameters will have to be specified. We illustrate this in Example 3 below.

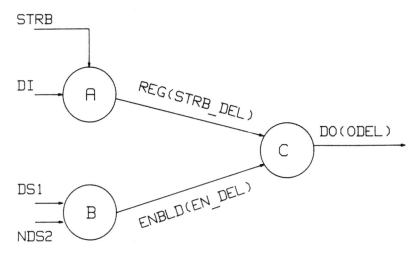

Figure 4.10 Process model graph for the register.

In light of the discussion above, the VHDL in Figure 4.11 is largely self-explanatory. However, note that in process C, the output process, we represent the high-impedance condition for the output as a vector of all 1's, as we have not yet introduced the concept of multivalued logic.

```
entity REG is
generic(STRB_DEL,EN_DEL,ODEL: TIME);
  port (DI :  in BIT_VECTOR (1 to 8);
   STRB: in BIT;DS1:  in BIT;
   NDS2: in BIT; DO: out BIT_VECTOR(1 to 8);
end REG;
architecture NODAL of REG is

  signal REG: BIT_VECTOR(1 to 8);
  signal ENBLD: BIT;

begin
A: process(STRB)
    begin
    if (STRB = '1') then
    REG <=DI after STRB_DEL;
    end if;
    end process A;

B: process(DS1,NDS2)
    begin
    ENBLD <= DS1 and  not NDS2 after EN_DEL;
    end process B;
```

```
C: process(REG,ENBLD)
   begin
   if (ENBLD = '1') then
   DO <= REG after ODEL; else
   DO <= "11111111" after ODEL;
   end if;
   end process C;
end NODAL;
```

Figure 4.11   VHDL description of data path model.

As we have stated in Chapter 2, individual signal assignment statements
are processes; therefore, the same model structure can be expressed more com-
pactly. Given below is an alternative architecture of the data path model which
uses this approach:

```
architecture DATA_FLOW of REG is
begin
B: block(STRB = '1'and not STRB'STABLE)
   signal REG: BIT_VECTOR(1 to 8);
   signal ENBLD: BIT;
begin
REG <= guarded DI after STRB_DEL;   ---process A
ENBLD <= DS1 and not NDS2 after EN_DEL;  ----process B
DO <=REG after ODEL when ENBLD = '1'  ---process C
      else "11111111" after ODEL;
   end block B;
end DATA_FLOW;
```

This approach uses a block that is guarded by a 0-to-1 transition on the
strobe. Each of the three processes A, B, and C is implemented by a single
statement. The statement corresponding to process A is guarded and will be
executed only when the guard condition is TRUE. This form of the process model
is more concise, but the structure of the model is not as evident. Also, for more
complicated modeling situations, implementing a complicated I/O formula
within a process block is more effective.

Example 2

For our second example of chip–level modeling, consider the parallel-
to-serial converter shown in Figure 4.12. The circuit operates as follows: (1)
the register SR is loaded when the strobe rises; (2) the serial data is shifted
out under control of an internal oscillator; and (3) the logic inhibits the load-
ing of inputs while shifting occurs. In addition, there are input specifications:
setup and hold-time requirements on the parallel input PI versus the strobe

(STRB). Finally, there is an output specification dictating that the serial output SO switches ODEL time units after the internal clock makes a 0-to-1 transition.

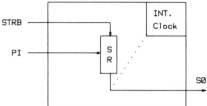

Figure 4.12 Parallel to serial converter.

Figure 4.13 shows a process model graph for the converter. The LOAD process loads data into the shift register and checks the setup and hold requirements. Assuming a proper load, the LOAD process signals the SHIFT process to begin via the SH signal. The shift register contents are stored in signal SR, which is accessed by both processes.When the shifting process is complete, the SHIFT process signals that fact by resetting SH. Signals SH and SR are signals with delta delay, so no delay values are indicated on the process model graph.

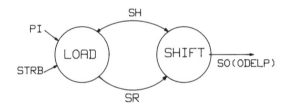

Figure 4.13 Process model graph for parallel to serial converter.

Figure 4.14 shows the VHDL model for the shifter. In the interface description, generic parameters are used for the timing and delay specifications. The LOAD process uses a transition on the strobe, delayed by an amount of time HT, to load the value of PI into the shift register, provided that PI has been stable for SUT + HT nanoseconds and a previous shift is not in process. If PI has not been stable for this period, the register SR will not be changed and an error will be reported. If the register is loaded, SH_1 is set to 1 to signal the shift process to begin.

The SHIFT process uses a "while" loop to implement the shifting process. Note that eight passes through the while loop are used to schedule completely the 8 serial output bits. When shifting is done, process SHIFT sets signal SH_2 to a "0." SH_1 and SH_2 are multiplexed together to form the signal SH. This multiplexing uses the 'QUIET attribute. For a signal X, X'QUIET is TRUE while no assignment is made to X. Thus for the multiplexer, the signal that is "not quiet" will determine the value of SH. This method of time multiplexing was developed by Dongal Han, a graduate research assistant in the Chip Level Modeling Research Group at Virginia Polytechnic Institute. We will discuss it further in Chapter 5.

```
entity PAR_SER is
generic(SUT,HT,PER,ODEL: TIME);
port(PI: in BIT_VECTOR(0 to 7); STRB: in BIT; SO: out BIT);
end PAR_SER;

architecture CHIP_LEVEL of PAR_SER is
signal SH_1,SH_2,SH: BIT;
signal SR: BIT_VECTOR(0 to 7);
begin
LOAD: process(STRB'DELAYED(HT))
        begin
        if (STRB'DELAYED(HT) = '1') then
        assert PI'STABLE(SUT + HT)
        report "Setup or Hold Time Failure";
        if PI'STABLE(SUT + HT) and SH = '0' then
          SR <= PI;
          SH_1 <= '1';
        end if;
        end if;
        end process LOAD;

SHIFT: process(SH)
        variable COUNT: INTEGER := 0;
        begin
        if (SH = '1') then
          COUNT := 0;
        while COUNT <8 loop
          SO <= transport SR(COUNT) after (COUNT*PER+ODEL);
          COUNT := COUNT + 1;
          end loop;
          SH_2 <= '0' after 8*PER;
        end if;
        end process SHIFT;
MUX: SH <= SH_1 when not SH_1'QUIET else
           SH_2 when not SH_2'QUIET else
           SH;

end CHIP_LEVEL;
```

Figure 4.14   VHDL description of the converter.

Example 3

In our concluding example we model a multimodule system that again illustrates the use of the process model graph and timing assertions. The example also illustrates the concept of functional partitioning that we described previously.

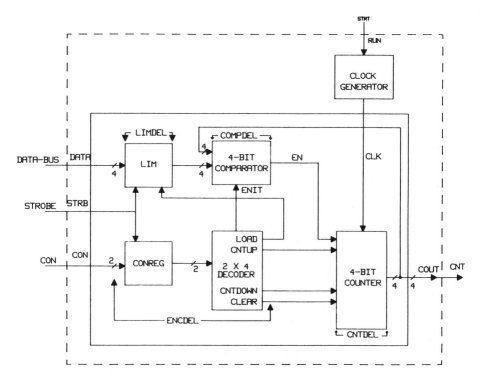

Figure 4.15   Clock generator and controlled counter.

In Figure 4.15 is shown the block diagram of a two-module system consisting of a clock generator and a 4-bit controlled counter. The clock generator produces a 50% duty cycle clock with a period of PER nanoseconds. The counter consists of five logic blocks: (1) a 2-bit control register (CONREG), (2) a 2-to-4 decoder, (3) a 4-bit limit register (LIM), (4) a 4-bit comparator, and (5) a 4-bit counter. The counter module receives a 2-bit control input (CON), which is stored in CONREG and then decoded to perform the following four commands:

1. (CON = 00). Clear the counter
2. (CON = 01). Load the LIM (limit) register with the value on the DATA input lines.
3. (CON = 10). Count up until the COUNTER equals the value in the LIM register.
4. (CON = 11). Count down until the COUNTER equals the value in the LIM register.

The counter, when commanded to count and enabled (EN = 1) counts positive clock transitions, in an up or down mode, until the limit is reached. When a counting command is decoded, the ENIT signal is sent to the comparator by the decoder logic. The comparator uses the rising edge of ENIT to set EN = 1 initially. After that, EN is controlled by the comparator.

The diagram shows four propagation delays: (1) ENCDEL—the combined delay across the control register (CONREG) and the decoder, (2) LIMDEL—the delay through the limit register, (3) COMPDEL—the delay through the comparator, and (4) CNTDEL—the delay through the counter. The interface timing functions as follows: CONREG is loaded on the rise of STRB, and LIM is loaded on the fall of STRB. There are setup-time requirements on both CON and DATA relative to their respective STRB transitions of TSET nanoseconds. There is a minimum pulse-width requirement of MPW on the STRB signal itself.

Figure 4.16 shows the process model graph for the controlled counter. Note that the functions of the control register (CON) and the decoder are represented by a single node. This is because the timing specification gives the total delay across the two blocks but not individually. Thus this node will be modeled with a single VHDL process. The counter node (CTR) has been functionally decomposed into CLR and a CNT section and thus will be modeled by two VHDL processes.

Figures 4.17 and 4.18 give the VHDL for the system described above. Note that there are two entities, one for the clock generator (Figure 4.17) and one for the 4-bit controlled counter (Figure 4.18).

In the clock generator model (Figure 4.17), the oscillator action is implemented by the feedback delay mechanism mechanized by a single process. The signal CLOCK, which is generated by the process, is one of the signals in the process sensitivity list. The output of the entity is CLK, which equals CLOCK after one delta delay. It is necessary to use this buffering mechanism since VHDL does not allow the output of an entity to be referenced within that entity. Note also that clock generation will occur only after the RUN input has made a 0-to-1 transition and will be inhibited after RUN falls.

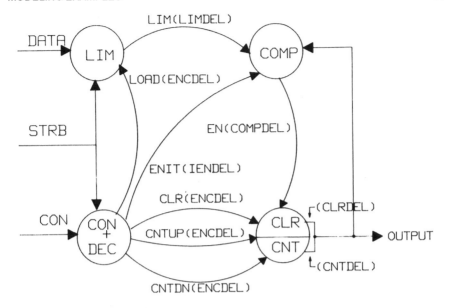

KEY	
CONSIG[0] ⇒ CLR	
CONSIG[1] ⇒ LOAD	
CONSIG[2] ⇒ CNTUP	
CONSIG[3] ⇒ CNTDWN	

Figure 4.16 Controlled counter process model graph.

The entity description for the controlled counter (Figure 4.18) first specifies a set of generics for the time delays in the model. Between the *begin* and *end* keywords the timing assertions are given. The architectural body contains five processes: DECODE, LOAD_LIMIT, CLEAR_CTR, CNT_UP_OR_DOWN, and LIMIT_CHK. Process DECODE decodes the 2-bit input CON into the 4-bit signal CONSIG. It also generates the signal ENIT1, which is used initially to enable the comparator. Processes LOAD_LIMIT and CLEAR_CTR are implemented as guarded blocks. Process COUNT_UP_OR_DOWN will respond to its clock input by counting only if the clock enable signal (CLKE) is set. CLKE is set by the rise of EN, the comparator enable signal, and reset by its fall. Process LIMIT_CHK sets the enable signal EN to 1 when ENIT is received and resets it to 0 when the count limit is reached. Finally, note that multiplexing is required to generate the signals CNT and ENIT. CNT is declared as a separate signal from COUT, because COUT cannot be read by a process.

```
entity CLOCK_GENERATOR is
 generic (PER: TIME);
 port(RUN: in BIT;CLK: out BIT);
end CLOCK_GENERATOR;
architecture IMPL_1 of CLOCK_GENERATOR is
 signal CLOCK: BIT;
 begin
  process (RUN,CLOCK)
  variable CLKE: BIT:= '0';
 begin
  if (RUN = '1' and not RUN'STABLE) then
   CLKE := '1';
   CLOCK <= transport '0' after PER/2;
   CLOCK <= transport '1' after PER;
  end if;
  if (RUN = '0' and not RUN'STABLE) then
   CLKE := '0';
  end if;
  if (CLOCK = '1'and not CLOCK'STABLE and CLKE = '1' ) then
   CLOCK <= transport '0' after PER/2;
   CLOCK <= transport '1' after PER;
  end if;
   CLK <= CLOCK;
  end process;
  end IMPL_1;
```

Figure 4.17 Clock generator description.

```
use work.COUNT_PAC.all;
entity CONTROLLED_CTR is
 generic(SUT,MPW,ENCDEL,IENDEL,CLRDEL,CNTDEL,LIMDEL,COMPDEL:
 TIME);
port(CLK,STRB: in BIT;
 CON:  in BIT_VECTOR(0 to 1);
 DATA  in BIT_VECTOR(0 to 3);
 COUT: out BIT_VECTOR(0 to 3));
begin
 assert STRB'STABLE or STRB = '0' or DATA'STABLE(SUT)
 report "Set up time failure on DATA input";
 assert STRB'STABLE or STRB = '1' or CON'STABLE(SUT)
 report "Set up time failure on CON input";
 assert STRB'STABLE or STRB = '1' or STRB'DELAYED'STABLE(MPW)
 report "Pulse width failure on STRB";

end CONTROLLED_CTR;
```

Figure 4.18 Counter description.

Architecture PROCESS_IMPL of CONTROLLED_CTR is

```
signal ENIT,ENIT1,ENIT2: BIT;
signal EN:    BIT;
signal CONSIG,LIM: BIT_VECTOR(0 to 3);
signal CNT,CNT1,CNT2: BIT_VECTOR(0 to 3);
begin

DECODE:

 process(STRB)
  variable CONREG: BIT_VECTOR(0 to 1) := '00'
 begin
  if (STRB = '1') then
   CONREG  := CON;
   case CONREG is
    when "00" =>  CONSIG <= "1000"
     after ENCDEL;
    when "01" =>  CONSIG <= "0100"
     after ENCDEL;
    when "10" =>  CONSIG <= "0010"
     after ENCDEL;
     ENIT1 <= '1' after IENDEL;
    when "11" =>  CONSIG <= "0001";
     after ENCDEL;
     ENIT1 <= '1' after IENDEL;
    end case;
   end if;

  end process DECODE;

LOAD_LIMIT:

 block(CONSIG(1)='1' and STRB ='0' and not STRB'STABLE)
  begin
   LIM <= guarded DATA after LIMDEL;
  end block LOAD_LIMIT;

CLEAR_CTR:

 block(CONSIG(0)='1' and not CONSIG(0)'STABLE)
 begin
  CNT1 <= guarded "0000" after CLRDEL;
 end block CLEAR_CTR;
```

Figure 4.18 Counter description. (continued)

```
CNT_UP_OR_DOWN:

process(CLK,EN)
 variable CLKE: BOOLEAN;

begin

 if  not EN'stable  then
  if EN = '1' then
   CLKE := TRUE;
  else
   CLKE := FALSE;
  end if;
 end if;

 if (CLK =  '1' and not CLK'STABLE and CLKE) then
  if (CONSIG(2)='1') then
   CNT2 <= INC(CNT);
  elsif (CONSIG(3)='1') then
   CNT2 <= DEC(CNT);
  end if;
 end if;

end process CNT_UP_OR_DOWN;

LIMIT_CHK:

   process(CNT,ENIT)
   begin
    if (ENIT = '1' and not ENIT'STABLE) then
     EN <= '1' after COMPDEL;
     ENIT2 <= '0' after COMPDEL;
    elsif ((EN = '1') and (CNT = LIM)) then
     EN <= '0' after COMPDEL;
    end if;
   end process LIMIT_CHK;

   MUX1: CNT <= CNT1 when not CNT1'QUIET else
                CNT2 when not CNT2'QUIET else
                CNT;
   MUX2: ENIT <= ENIT1 when not ENIT1'QUIET else
                 ENIT2 when not ENIT2'QUIET else
                 ENIT;
   COUT <= CNT;

 end PROCESS_IMPL;
```

Figure 4.18 Counter description. (continued)

The descriptions given in Figures 4.17 and 4.18 specify the clock generator and the controlled counter as generic design entities. To implement the complete system shown in Figure 4.15, one has to specify the connection between the two design entities. This is done in Figure 4.19. First an entity is defined for the total system. In Figure 4.15 this system entity is shown inside the dashed box. The port statement for this entity defines the signals that cross the system boundary. Note that the names for signals can be different at the system boundaries and internal to the system. The architectural body is structural. First the two components and an internal signal between them (CLK) are declared. Between the *begin* and *end* keywords the two components are instantiated; that is, generic maps and port maps are used to give values to the generic parameters and connect signals to the correct inputs.

```
entity COUNT_SYS is
port(STRT,STROBE: in BIT;
  CON: in BIT_VECTOR(0 to 1);
  DATA_BUS: in BIT_VECTOR(0 to 3);
  CNT: out BIT_VECTOR(0 to 3));
end COUNT_SYS;

architecture TWO_COMPONENT of COUNT_SYS is

signal CLK: BIT;
component CLOCK_GENERATOR
generic(PER: TIME);
port (RUN: in BIT;
      CLK: out BIT);
end component;
component CONTROLLED_CTR
generic(SUT,MPW,ENCDEL,IENDEL,CLRDEL,CNTDEL,LIMDEL,
        COMPDEL: TIME);
port (CLK,STRB: in BIT;
      CON: in BIT_VECTOR(0 to 1);
      DATA: in BIT_VECTOR(0 to 3);
      COUT: out BIT_VECTOR(0 to 3));
end component;

begin

CLKGEN:
  CLOCK_GENERATOR
    generic map(100 ns)
    port map(STRT,CLK);
CTR:
  CONTROLLED_CTR
    generic map(20 ns,30 ns,25 ns,20 ns,
            10 ns,15 ns,12 ns,10 ns);
    port map(CLK,STROBE,CON,DATA_BUS,CNT);

end TWO_COMPONENT;
```

Figure 4.19 Interconnected counter system.

SUMMARY

In this chapter we have defined what a chip-level model is and have shown that there are two main elements of this model. First, the model structure can be represented by a process model graph. Second, the modeling of accurate timing depends on the inertial delay mechanism in the basic VHDL "after" clause and the use of timing assertions. Three levels of timing modeling were presented: delta delay timing, simple I/O timing, and data path timing. Only data path timing fully meets the chip-level model timing accuracy requirement. The three examples in the chapter reinforced these concepts.

BIBLIOGRAPHY

Armstrong, J. R., "Chip Level Modeling and Simulation," *Simulation*, pp. 141-148, October 1983.

Armstrong, J. R., "Chip Level Modeling of LSI Devices," *IEEE Transactions on Computer-Aided Design*, Vol. CAD-3, No. 4, pp. 288-297, October 1984.

Armstrong, J. R., "Chip Level Modeling With HDLs," *IEEE Design and Test of Computers*, pp. 8-18, February 1988.

IEEE Computer, "Special Issue on Hardware Description Languages," February 1985.

5

SYSTEM MODELING

In our discussion of system modeling, we consider two main aspects of this problem: system interconnection schemes and the modeling of the generic component types used in today's computer systems: processors, memories, UARTS, and parallel ports. The generic component modeling will be illustrated by means of a comprehensive system example.

THE MODELING OF SYSTEM INTERCONNECTION

In this section we consider the problem of representing signal interconnect between components. We first present a general model for signal interconnection. Next we discuss bundled signal representation and the use of multiple-valued logic. Finally, we explain how the multiplexing of signals is performed at the system level.

GENERAL MODEL FOR SIGNAL INTERCONNECTION

In VHDL the general model for signal interconnection is implemented in a structural architectural body representing the system entity. Let us illustrate this by means of an example. Suppose that a system X consists of two components A and B, and that A and B have the following interface descriptions:

```
entity A is
  port(P: in BIT; Q: out BIT);
end A;

entity B is
  port(R: in BIT; S: out BIT);
end B;
```

Furthermore, suppose that A and B are interconnected as shown in Figure 5.1 to form system X. Then X can be represented by the following entity declaration and accompanying architectural body:

```
entity X  is
  port(L: in BIT; M: out BIT);
end X;

architecture STRUCTURAL of X is
  signal S1: BIT;              ----- Signal declared
  component A                  ----- Component declared
    port(P: in BIT; Q: out BIT);
  end component;
  component B                  ----- Component declared
    port(R: in BIT; S: out BIT);
  end component;
begin
  A1: A port map(L,S1);        ---- Component Instantiation
  B1: B port map(S1,M);        ---- Component Instantiation
end STRUCTURAL;
```

Thus the interconnection of system components is modeled by signals between components which are first declared and then instantiated. In the simple case shown above, the signal S1 comprises the system interconnection. For more complicated examples, the reader is referred to the example of the controlled

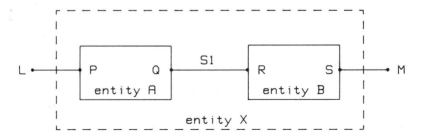

Figure 5.1   System interconnection model.

counter in Chapter 4 (Figure 4.19) and the comprehensive system example (Figure 5.7) of this chapter.

Note in the example above that when the components are instantiated, the interconnect is specified by position matching of port map entries in the instantiation statements with port entries in the component declarations. This can also be performed by named association. Thus in VHDL one could also write for the instantiation statements:

```
A1: A port map(Q => S1, P => L);
B1: B port map(S => M , R => S1);
```

Note here that position is unimportant and that the named association identifies the correspondence.

Bundled Signal Representation

A designer may want to vary the level of view when modeling a system. Consider the following example, shown in Figure 5.2. A designer wishes to model the interface between a control unit (CU) and an arithmetic unit (AU). The CU sends a 5-bit code to the AU, specifying the operation to be performed. Shown below is a definition of this interface. The two components CU and AU are assumed to be part of system SYS, but for sake of clarity, only the description related to those two components is shown:

```
entity SYS is
   port(............ ); ----- define external system inputs here.
end SYS;

architecture INTERCONNECT of SYS
   signal C : BIT_VECTOR(4 downto 0);
   component  CU
      port(      ; COUT: out BIT_VECTOR(4 downto 0));
   end component;
   component AU
      port(      ; AUIN: in   BIT_VECTOR(4 downto 0));
   end component;
   begin
   CONTROL_UNIT: CU portmap(...... ; COUT => C);
   ARITH_UNIT: AU portmap(...... ; AUIN => C);
   end INTERCONNECT;
```

The definition above models the interface at the bit level. Ultimately, this interface would have to be modeled at this level, but initially the designer may wish to model abstractly, not yet making specific code assignments.

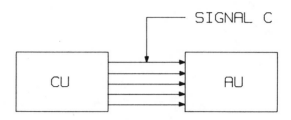

Figure 5.2   Control unit interface.

Simulation efficiency is another consideration. The control interface between the CU and the AU is essentially a logic entity. Yet if modeled at the bit level, the simulator will have to process events on the individual lines; for example, if all five lines change at the same time, five events must be processed. If the lines could somehow be bundled together and regarded as an entity, only one event would need to be processed.

Use of Data Types

To model the signal interface at a high level, one can use other data types to represent the information. For example, instead of declaring C as

```
signal C: BIT_VECTOR(4 downto 0);
```

one could make the following definition:

```
type CONTROL is   (AND,OR,XOR,ADD,SUB,........,TCOMP);
signal    C: CONTROL;
```

Thus using an enumeration type, the control information is represented in mnemonic form. Using this form, assignment statements such as

```
COUT<= ADD;
```

could be made in the control unit architectural body, while in the arithmetic unit body one could write

```
if (AUIN = ADD) then etc.
```

The modeler is thus removed from worrying about low-level details.
Another possible definition for the interface is

```
subtype   CONTROL is INTEGER range 0 to 31;
signal    C: CONTROL;
```

Here the designer would use integers to represent the control codes, and these integer values would not necessarily be the numerical values finally assigned in the design. Type INTEGER could also be used, but the use of subtype CONTROL restricts the value that can be assigned to C, thus providing error checking.

Even though an interface is to be modeled as an integer type, in some cases it may be desirable to work at the bit level inside the models. To do this, one must define type conversion functions between type BIT_VECTOR and type INTEGER. Figure 5.3 shows two such functions. Function BIN5_TO_INT converts bit vectors to integers by summing appropriate powers of 2. Function INT_TO_BIN5 performs the inverse operation by division of successive powers of 2.

```
function BIN5_TO_INT
(INPUT: BIT_VECTOR(4 downto 0))
return INTEGER is
variable SUM   : INTEGER := 0;
begin
for I in 0 to 4 loop
  if (INPUT(I) = '1') then
    SUM := SUM + (2**I);
  end if;
 end loop;
 return SUM;
end BIN5_TO_INT;

function INT_TO_BIN5
(INPUT : INTEGER)
return BIT_VECTOR(4 downto 0) is
variable FOUT: BIT_VECTOR(4 downto 0);
variable TEMP_A: INTEGER:= 0;
variable TEMP_B: INTEGER:= 0;
begin
 TEMP_A := INPUT;
 for I in 4 downto 0 loop
  TEMP_B := TEMP_A/(2**I);
  TEMP_A := TEMP_A rem (2**I);
  if (TEMP_B = 1) then
   FOUT(I) := '1'; else
   FOUT(I) := '0';
  end if;
 end loop;
 return FOUT;
end INT_TO_BIN5;
```

Figure 5.3   Conversion functions.

The interconnection between the two system blocks could then be specified as follows:

```
entity SYS is
(                    )
end SYS;
architecture INTERCONNECT of SYS
 signal C : INTEGER;
 component  CU
   port(........; COUT: out BIT_VECTOR(4 downto 0));
 end component;
 component AU
   port(.......; AUIN: in   BIT_VECTOR(4 downto 0));
 end component;
begin
 CONTROL_UNIT: CU port map(........; BIN5_TO_INT(COUT)=>C);
 ARITH_UNIT: AU port map(........ ; AUIN => INT_TO_BIN5(C));
end INTERCONNECT;
```

In the case above, the type conversions have been incorporated directly into the component instantiation statements. One could also, of course, do the type conversions internal to the architectural bodies for the CU and the ALU, but this is a less general approach.

MULTIPLEXING OF SIGNALS

In this section we consider various approaches to signal multiplexing. The approaches can be divided into two types: the use of bused signals and time multiplexing.

Bused Signals

We define a bused signal to be one that is multiply driven. Recall that in Chapter 2 we discussed the concept of a signal having multiple drivers. Consider again the situation shown in Figure 2.11. Both process A and process B make assignments to the signal X. In VHDL this causes two "drivers" of the signal, Dax and Dbx, to be created. The signal X is then a function of both its drivers. The function in this case is F; in VHDL this situation can be declared as follows:

```
type MVL is ('0','1','Z');
signal  X: F MVL;
```

X is declared to be a signal of type MVL. Listing "F" specifies the function that is used to resolve the values of the multiple drivers of X into the single value of the signal X. F is a function that must have been declared previously and is called whenever one of the drivers of X change. It is left to the user to define the function F, which is application and technology dependent. One possible implementation of F for the tristate logic system declared above is

```
function F(signal S: TSV)
  return MVL is
  variable RESOLVED_VALUE: MVL:='Z';
begin
  for I in S'RANGE loop
    if S(I) /= 'Z'  then
      RESOLVED_VALUE := S(I);
      exit;
    end if;
  end loop;
  return RESOLVED_VALUE;
end F;
```

Note that the function above scans the drivers of S and selects the value of the first driver that is not equal to Z, the assumption being that only one of the drivers is active. The drivers of S are implicitly regarded as an array and the 'RANGE attribute is used to specify the limits of this array. The drivers of S are of type TSV (tristate vector), while the result of the function is of type MVL ('0','1','Z'). More will be said about these types when we discuss multivalued logic.

The assumption in the function above that only one driver is active is certainly not valid in all modeling situations. However, other implementations are possible; for example, in order to implement a wired-or resolution function, one could replace the loop in the function F with the following:

```
for I in S'RANGE loop
  if S(I) = '1' then
    RESOLVED_VALUE = '1';
    exit;
  elsif S(I) = '0' then
    RESOLVED_VALUE = '0';
  end if;
end loop;
```

Note in this function that detection of any 1's will result in a 1 being returned and that both 1's and 0's predominate over Z. The wired AND function can be defined similarly but is left as an exercise for the reader.

Time Multiplexing

Guarded blocks.   An interesting situation develops when the busing of signals is combined with guarded signal assignment statements. Consider the following two VHDL blocks:

```
- - - - - - - - - - - - - - - - - - - - -
- - - - - - - - - - - - - - - - - - - - -
type  MVL  is  ('0','1','Z');
signal   X:  F  MVL;
begin
 ONE:  block(PHASE_ONE)
 begin
  X <= guarded DB1    after DEL1;
 end block ONE;

 TWO:  block(PHASE_TWO)
 begin
  X <= guarded DB2    after DEL2;
 end block TWO;
 - - - - - - - - - - - - - - - - - - - - -
 - - - - - - - - - - - - - - - - - - - - -
```

Assume that in the logic to be modeled, PHASE_ONE and PHASE_TWO are mutually exclusive outputs of a cycle counter. The intent of the model is that when PHASE_ONE becomes TRUE, DB1 will be assigned to X, while when PHASE_TWO becomes TRUE, DB2 will be assigned to X. However, with the definition of busing above, this will not happen. The signal X is driven by two process statements and thus has two drivers. Whenever X is referenced, the value of X that is used is a static function of both drivers. However, this was not the intent. The intent was to have X be time multiplexed, that is, be a function only of the driver from the block whose guard expression is TRUE.

To handle this problem of bus resolution combined with guarded signal assignment statements, VHDL provides special mechanisms for signals used in guarded signal assignments. If the modeler wishes to imply time multiplexing, the signals are designated as being a "register" or "bus"; for example, in the example above, X would be declared as

signal X: F MVL register;

or

signal X: F MVL bus;

If a guarded signal is designated as being a register or a bus, and if the guarded statement driving it has its block guard become FALSE, the statement's driver is assumed to be off (i.e., it is ignored by the bus resolution function). Thus in the case shown above, the block whose guard is TRUE will have its value come through the bus resolution function. For the case where all the block guards are off, if X is a "register," it will retain its last value; if X is a "bus," it will be assigned the default value provided by the bus resolution function. For the bus resolution functions defined above, the value of Z is returned as a default value if the function has no active inputs. In summary, if X is a register, clocked register multiplexing is implied. If X is a bus, clocked bussing is implied. The equivalent logic for these two situations is shown in Figure 5.4 a and b.

It should be emphasized that these designations "register" and "bus" are used in VHDL only with guarded signals. Thus a signal that is multiply driven implies a bus resolution function, but it cannot be declared to be a "bus" or "register" unless it is the object of a guarded signal assignment. The reader should keep this restrictive use of the terminology "register" and "bus" in VHDL in mind.

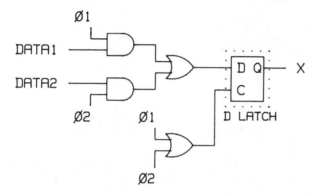

Figure 5.4(a)   Clocked register multiplexing.

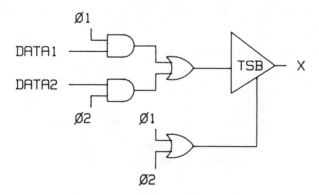

Figure 5.4(b)   Clocked bus multiplexing.

Processes.   A similar time-multiplexing problem can develop with processes. Consider the following simple example:

```
A: process(SET)          B: process(RESET)
     begin                     begin
       if SET='1' then           if RESET='1' then
         X <='1';                  X <='0';
       end if;                   end if;
   end process A;            end process B;
```

The intent of the two processes is to have process A set X on the rise of SET, and process B reset X on the rise of RESET. This is again a time-multiplexing problem (i.e., a bus resolution function alone cannot model this behavior). We offer two solutions to this problem. The first solution is to enclose the processes inside guarded blocks:

```
A: block(SET='1')        B: block(RESET='1')
     begin                     begin
       process(GUARD)            process(GUARD)
       begin                     begin
         if GUARD then             if GUARD then
           X <= '1';                 X <= '0';
         else                      else
           X <= null;                X <= null;
         end if;                   end if;
       end process;              end process;
   end block A;              end block B;
```

Note now that for either process, the driver for X from that process will receive "null" when GUARD is FALSE, thus disconnecting that driver from the bus resolution function defined for X. GUARD is a signal implicitly defined within any guarded block whose value is the same as the guard expression.

Although effective, the approach above makes model descriptions verbose, particularly when a signal has a number of assignments made to it from different processes. Perhaps a better approach is to dispense with bus resolution functions entirely and define multiplexers external to the processes:

```
A: process(SET)          B: process(RESET)
     begin                     begin
       if SET='1'then            if RESET='1'then
         X1 <= '1';                X2 <='0';
       end if;                   end if;
   end process A;            end process B;

         X <= X1 when not X1'QUIET else
              X2 when not X2'QUIET else
              X;
```

Note that processes A and B now assign values to distinct signals X1 and X2, which are then multiplexed. Recall that this form of multiplexing was used in two models in Chapter 4. The penalty for this approach is that extra signals and the multiplexer must be defined; however, the overall impact on the size of the description is less than with the use of the guarded blocks. Also, with this second approach, the process descriptions are less cluttered.

It is unfortunate that one has to add these special mechanisms to perform time multiplexing with VHDL processes. It is one of the weaknesses in the current version of the language that hopefully future releases will correct.

MULTIPLE-VALUED LOGIC

While boolean functions operate on the binary values 0 and 1, the modeling of real logic circuits frequently requires the use of more than two values (i.e., multiple-valued logic). Above, when discussing bus resolution functions, we considered briefly the case of a three-valued system {0,1,Z}. Other possibilities include {0,1,X,Z,U,D}, that is, a six-valued system. Systems of this type are frequently employed to model gate-level logic. In fact, as many as 15 values have been used for this purpose.

When modeling at the chip level, multiple-valued logic is used primarily to describe interface phenomena. This is because it is difficult to propagate the effect of nonbinary values through behavioral models. For example, suppose that a behavioral model contains the following statement:

```
case A is
  when '0'=> --action for logic 0;
  when '1'=> --action for logic 1;
  when 'X'=> --action for logic X;
  when 'Z'=> --action for logic Z;
end case;
```

Then the question is: What action should be taken if A takes on the values of Z (high impedance) or X (indeterminate)? The Z case can usually be resolved by technology considerations; for example, for the case of TTL or MOS, one could reasonably assume that Z is a logic 1. For the case of A= 'X', however, the situation is much more difficult. One possible approach is to assume randomly that A is 0 or 1, but this results in a nondeterministic simulation. Another approach is to "spawn" a new simulation for each possible value for A, but this approach rapidly becomes too difficult to evaluate. The situation above demonstrates the effect that multiple-valued signals can have on the control flow of behavioral models. Similar problems are involved in defining multiple-valued data operations for behavioral models, although they may be more tractable than the control case. In summary, the use of multiple-valued logic internal to behavioral models remains an open research problem. We thus restrict our attention to the modeling of signal interfaces between models, and for our system example, the three-valued case {0,1,Z}.

To model a multiple-valued logic interface, one must define:

1. A multivalued logic type
2. A multivalued array type
3. Type conversion functions for individual signals
4. Type conversion functions for arrays of signals
5. A bus resolution function

Figure 5.5 gives a complete package to implement a three-valued logic system. The general type MVL is first declared as a three-valued enumeration type. Next, an array type TVECT is defined. This type is used only by the bus resolution function, BUS FUNC, which is defined next. Following this, a subtype of MVL, TSL, is defined which employs the resolution function BUSFUNC. All signals of the subtype TSL will automatically have BUSFUNC as their bus resolution function. The array, or vector, type TSV is next defined as a composite type whose individual elements are of subtype TSL. The range of the type is not specifically given at this point but will be at the point of its use.

The function TRISTATE_TO_BIT converts type TSL to type BIT. When the input value is equal to Z, the value returned is the constant 'TIE_OFF', which is specified external to the function and is given a value related to the technology employed. The function BIT_TO_TRISTATE performs the opposite conversion. It is interesting that this function has as its input the possible values '0' and '1'(the range of indices of the array type TSA are the members of type BIT), and its output apparently has the same values. However, the values '0' and '1' in question belong to two different data types, and thus the type conversion is necessary. The conversion is implemented by accessing a linear array of length 2.

The functions BITVEC_TO_TRIVEC and TRIVEC_TO_BITVEC perform the vector conversions by applying the single-line functions to each vector position. The range of the conversion loop is specified by using the attribute 'RANGE of the signal INPUT. Finally, the bus resolution function is given. Recall that this function is used to derive the value of a single line that is multiply driven. This line may be a single signal, or it may be part of an array of signals. All the drivers of a given signal are implicitly regarded as an array, and thus BUSFUNC examines signals in INPUT'RANGE of the array of signal drivers. The rationale for the operation of this bus function was given above.

```
package TSL is

  type MVL is ('0','1','Z');
  type TVECT is array (INTEGER RANGE <>) of MVL;
  function BUSFUNC(INPUT: TVECT) return MVL;
  subtype TSL is BUSFUNC MVL;
  type TSV  is array (INTEGER RANGE <>) of TSL;
  constant TIE_OFF: BIT:='1';
  function TRISTATE_TO_BIT (INPUT: TSL) return BIT;
  function BIT_TO_TRISTATE (INPUT: BIT) return TSL;
  function BITVEC_TO_TRIVEC (INPUT: BIT_VECTOR) return TSV;
  function TRIVEC_TO_BITVEC ( INPUT: TSV) return BIT_VECTOR;

end TSL;

package body TSL is

  function TRISTATE_TO_BIT (INPUT: TSL) return BIT is
  begin
   case INPUT is
    when 'Z' => return TIE_OFF;
    when '0' => return '0';
    when '1' => return '1';
   end case;
  end TRISTATE_TO_BIT;

  function BIT_TO_TRISTATE (INPUT: BIT) return TSL  is
   type TSA is array(BIT) of MVL;
   constant TSL_VALUE: TSA :="01";
  begin
   return TSL_VALUE(INPUT);
  end BIT_TO_TRISTATE;

  function BITVEC_TO_TRIVEC (INPUT: BIT_VECTOR)
   return TSV  is
   variable CONVERTED_VALUE: TSV(INPUT'RANGE);
  begin
   for I in INPUT'RANGE loop
    CONVERTED_VALUE(I) := BIT_TO_TRISTATE (INPUT(I));
   end loop;
   return CONVERTED_VALUE;
  end BITVEC_TO_TRIVEC;
```

Figure 5.5 Tristate logic package.

```
function TRIVEC_TO_BITVEC (INPUT: TSV)
  return BIT_VECTOR   is
  variable CONVERTED_VALUE: BIT_VECTOR(INPUT'RANGE);
begin
  for I in INPUT'RANGE   loop
    CONVERTED_VALUE(I) := TRISTATE_TO_BIT (INPUT(I));
  end loop;
  return CONVERTED_VALUE;
end TRIVEC_TO_BITVEC;

function BUSFUNC(INPUT: TVECT)
  return MVL is
  variable RESOLVED_VALUE: MVL:='Z';
begin
  for I in INPUT'RANGE loop
   if INPUT(I) /= 'Z'   then
     RESOLVED_VALUE := INPUT(I);
     exit;
   end if;
  end loop;
  return RESOLVED_VALUE;
end BUSFUNC;

end TSL;
```

Figure 5.5   Tristate logic package. (continued)

COMPREHENSIVE SYSTEM EXAMPLE

In this section we give a comprehensive system modeling example. The model consists of a computer system with the following components:

1. A small processor
2. A RAM memory
3. A parallel input port
4. A parallel output port
5. A parallel–serial/serial–parallel converter (UART)
6. An interrupt controller

Although its real capability is modest, the system illustrates the modeling of generic types of logic.

The block diagram of the system is shown in Figure 5.6. We first discuss the system as a whole and then analyze each component model in detail. In this detailed component analysis, we also describe variations on the basic model structure.

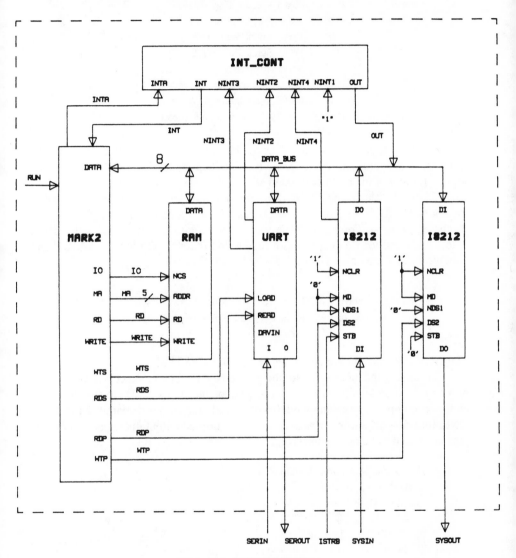

Figure 5.6   MARK2 SYSTEM.

The address space of the system is of size 32. The low 24 locations are assigned to the RAM memory and the 8 highest locations are used to implement a memory-mapped I/O scheme and interrupt structure. The assignment of addresses for this scheme is as follows:

11000	Read Parallel Port (RDP)
11001	Write Parallel Port (WTP)
11010	Read Serial Port (RDS)
11011	Write Serial Port (WTS)
11100	spare
11101	Serial Output Interrupt Vector
11110	Serial Input Interrupt Vector
11111	Parallel Input Interrupt Vector

The first four addresses are decoded internal to the processor to generate the I/O interface signals shown in Figure 5.6. Another signal, labeled "IO," will be a logic 1 for any of the addresses listed above and zero otherwise.

Figure 5.7 shows the VHDL description of the system interconnect. Note that the entity MARK2_SYS uses the packages TSL and SYSTEM. Package TSL was discussed in the section on multiple-valued logic. Package SYSTEM is discussed below in the section on the processor model. In Figure 5.7 the subtype WORD and subtype ADDR are used. They are declared as follows:

```
subtype WORD is TSV(7 downto 0);
subtype ADDR  is BIT_VECTOR(4 downto 0);
```

These two declarations are contained in package SYSTEM.

Following the method illustrated previously, the interconnect is implemented in a structural architectural body where components and signals are first declared and then components instantiated. Two special signals are declared: ZERO and ONE, for tying off unused inputs. Note that an unused output in the parallel output port is marked as being "open" (i.e., unconnected).

```
use work.TSL.all, work.SYSTEM.all;
entity MARK2_SYS is
port(SYSOUT : out  WORD;
     SYSIN    : in    WORD;
     ISTRB    : in    BIT;
     SEROUT  : out  TSL;
     SERIN    : in    TSL;
     RUN      : in    BIT);
end MARK2_SYS;
```

Figure 5.7 System interconnect.

```
architecture   CHIP_LEVEL of MARK2_SYS        is

    component   MARK2
                generic(RDEL,WDEL,ODEL,MADEL,INTDEL,PER: TIME);
                port (DATA: inout WORD;
                      MA: out ADDR;
                      RUN,INT: in BIT;
                      RD,WRITE,IO,RDS,RDP,WTS,WTP,
                      INTA: out BIT);

    end component;
    component   RAM
                generic(RDEL,DIS: TIME);
                port (DATA: inout WORD;
                      ADDR: in ADDR;
                      RD,WRITE,NCS: in BIT);

    end component;
    component   UART
                generic(CLK_PER,ODEL,INDEL,INTDEL: TIME);
                port (DATA: inout WORD; I:in TSL;
                      LOAD,READ: in BIT; O:out TSL;
                      NINTO,NINTI: out BIT);

    end component;
    component   I8212
                generic(GDEL,FFDEL,BUFDEL: TIME);
                port (DI: in   WORD;
                      DO: out WORD;
                      NDS1,DS2,MD,STB,NCLR: in BIT;
                      NINT: out BIT);

    end component;
    component INT_CON
                generic(INTDEL,ENCDEL,BUFFDEL: TIME);
                port (NINT1,NINT2,NINT3,NINT4,INTA: in BIT;
                                   INT: out BIT; VOUT: out WORD);

    end component;
    signal DATA_BUS : WORD;
    signal MA : ADDR;
    signal RD, WRITE, RDS, RDP: BIT;
    signal WTS, WTP, IO: BIT;
    signal INT,INTA,NINT1,NINT2,NINT3: BIT;
    signal ZERO: BIT := '0';
    signal ONE: BIT   := '1';
```

Figure 5.7   System interconnect. (continued)

```
begin

   CPU:   MARK2
          generic map(150 ns,100 ns,50 ns,40 ns,150 ns,500 ns)
          port map(DATA_BUS, MA, RUN, INT, RD, WRITE, IO,
                   RDS, RDP, WTS, WTP, INTA);

   MEM: RAM
          generic map(100 ns,30 ns)
          port map(DATA_BUS, MA, RD,
                   WRITE, IO);

   SER:   UART
          generic map(104 us,20 ns,30 ns,30 ns)
          port map(DATA_BUS, SERIN, WTS, RDS,SEROUT,
                   NINT3,NINT2);

   PARIN:  I8212
          generic map(20 ns,30 ns,20 ns)
          port map(SYSIN, DATA_BUS, ZERO, RDP, ZERO, ISTRB,
                   ONE,NINT1);

   PAROUT: I8212
          generic map(20 ns,30 ns,20 ns)
          port map(DATA_BUS, SYSOUT, ZERO, WTP, ONE, ZERO,
                   ONE,open);
   SYSINT: INT_CON
          generic map(20 ns,30 ns,20 ns)
          port map(NINT1,NINT2,NINT3,ONE,INTA,INT,DATA_BUS);

end CHIP_LEVEL;
```

Figure 5.7   System interconnect. (continued)

PROCESSOR MODEL

For the processor model, we choose a processor that has a small instruction set size but still contains the essential elements of most general-purpose computers. The processor unit, designated the MARK2, is similar to the 1940s processor MARK1, which could execute only seven instructions. However, the MARK2 model shown here implements a somewhat different set of arithmetic and control instructions. Also, the following features have been added to give the processor a more modern look: (1) a primitive interrupt capability which requires an instruction for enabling, disabling, and returning from interrupts—thus our processor model has eight instructions, (2) control signals for external memory, and (3) logic to implement the memory-mapped I/O scheme discussed previously.

The word lengh of the MARK2 is 8 bits, which allows for the address length of 5 bits and instruction opcode length of 3 bits. Thus the instruction format is as follows:

```
7   6   5     4   3   2   1   0

- - - - - - - - - - - - - - - - - - - - - - - - - - - - - - -
|  op  code  |          address              |
- - - - - - - - - - - - - - - - - - - - - - - - - - - - - - -
```

The descriptions of the eight instructions themselves are as follows:

Instruction	Opcode	Description
JMP	000	absolute jump;
TCA	001	take the twos complement of the accumulator;
LDA	010	load the accumulator;
STA	011	store the contents of the accumulator;
ADD	100	add the addressed operand to the accumulator;
INT	101	interrupt control;
JPN	110	jump if accumulator negative;
STP	111	stop;

The interrupt control instruction (INT) microcodes the address to perform three functions: (1) address bit 4 set—enable interrupts, (2) address bit 3 set—disable interrupts, or (3) Address bit 2 set—return from interrupt. The MARK2 model will be used to illustrate a general processor model structure. The process model graph for the MARK2 is shown in Figure 5.8. The model consists of the following components:

1. A process used to generate the processor clock (CLK)
2. A RUN process used to control stopping and starting of the model
3. A STATE process to initiate entrance into the FETCH, EXECUTE, and INTERRUPT processes
4. A FETCH process
5. An EXECUTE process
6. An INTERRUPT process
7. Data flow logic to decode the I/O control signals and multiplex process outputs

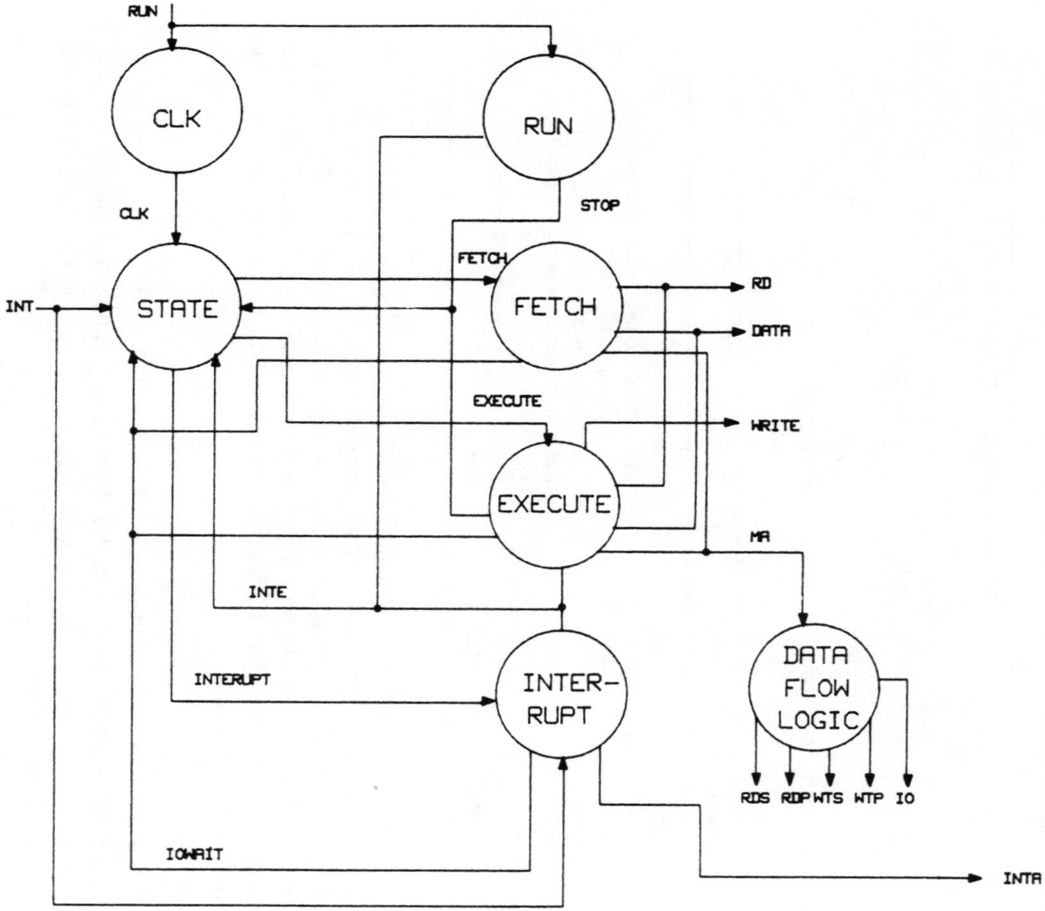

Figure 5.8 Processor model graph.

In addition to these main processor model components, it is necessary to develop a package that contains arithmetic functions and data types. One function that is required is an 8-bit add. This functions is required if one does not use type INTEGER representation of numbers.

As explained above, for convenience sake, we define subtype WORD to be TSV(7 downto 0). The reason that WORD is a tristate vector instead of type BIT_VECTOR is that we wish to avoid frequent type conversions at the system interface.

Figure 5.9 shows the package SYSTEM, which contains the subtype definitions WORD and ADDR, as well as the above-mentioned add function. In addition, the package contains:

1. Two functions (INC_WORD and INC_ADDR) for incrementing signals and variables of subtypes WORD and ADDR.
2. An invert function (INVW) for signals and variables of subtype WORD.
3. A function to convert subtype ADDR to type INTEGER (INTVAL). This is used in the RAM memory model.
4. A type conversion function for converting subtype WORD to subtype ADDR (VDAD). This is used in process INTERRUPT.

```
package SYSTEM is

    subtype WORD is TSV(7 downto 0);
    subtype ADDR is BIT_VECTOR(4 downto 0);
    function ADD8( A: WORD; B: WORD)  return WORD;
    function INC_WORD( COUNT: WORD)    return WORD;
    function INC_ADDR( COUNT: ADDR)    return ADDR;
    function INVW( X: WORD) return WORD;
    function INTVAL( VAL: ADDR) return INTEGER;
    function VDAD( DATA: WORD) return ADDR;

end SYSTEM;
```

Figure 5.9   Package SYSTEM.

```
package body SYSTEM is

function ADD8( A: WORD; B: WORD)
return WORD     is
variable CARRY: MVL := '0';
variable S: TSV(1 to 3);
variable NUM: INTEGER range 0 to 3 := 0;
variable SUM: WORD;
begin
 for I in 0 to 7 loop
 S:=A(I) & B(I) & CARRY;
  for K in 1 to 3 loop
   if S(K)='1'  then
   NUM := NUM + 1;
   end if;
  end loop;
  case NUM is
   when 0 => SUM(I) := '0'; CARRY := '0';
   when 1 => SUM(I) := '1'; CARRY := '0';
   when 2 => SUM(I) := '0'; CARRY := '1';
   when 3 => SUM(I) := '1'; CARRY := '1';
  end case;
  NUM := 0;
 end loop;
 return SUM;
end ADD8;

function INC_WORD( COUNT: WORD)
return WORD is
 variable A: WORD;
begin
 A:= COUNT;
 for I in COUNT'LOW to COUNT'HIGH loop
  if A(I) = '0' then
   A(I) := '1';
   exit;
  else A(I) := '0';
  end if;
 end loop;
 return A;
end INC_WORD;
```

Figure 5.9 Package SYSTEM. (continued)

```
--------- define INC_ADD similarly

function INVW( X: WORD)
 return WORD is
 variable TEMP: WORD;
begin
 for I in 0 to 7 loop
  if X(I) = '0' then
   TEMP(I) := '1';
  elsif X(I) = '1' then
   TEMP(I) := '0';
  else
   assert X(I) /= 'Z'
   report "Internal Data = to Z, set = to 1" ;
   TEMP(I) := '1';
  end if;
 end loop;
 return TEMP;
end INVW;


function INTVAL( VAL: ADDR)
 return INTEGER  is
 variable SUM: INTEGER := 0;
begin
 for N in VAL'LOW  to  VAL'HIGH  loop
  if VAL(N) = '1'  then
  SUM := SUM + (2 ** N);
  end if;
 end loop;
 return SUM;
end INTVAL;
```

Figure 5.9 Package SYSTEM. (continued)

```
function VDAD( DATA: WORD)
return ADDR is
variable ADDRESS: ADDR;
begin
for I in 4 downto 0 loop
 if DATA(I) = '0'then
  ADDRESS(I) = '0';
  elsif  DATA(I) = '1' then
  ADDRESS(I) = '1';
  else ADDRESS(I) = '1';
  endif;
  assert DATA(I) /= 'Z'
  report "Improper Int. Vector";
 end loop;
 return ADDRESS;
end VDAD;

end SYSTEM;
```

Figure 5.9 Package SYSTEM. (continued)

The next step in modeling the MARK2 is its definition as a design entity. Figure 5.10 shows this. Note that six generic time parameters are declared (RDEL, WDEL, ODEL, MADEL, INTDEL, PER). Their use is explained below.

```
use work.TSL.all, work.SYSTEM.all;
entity  MARK2 is
 generic(RDEL,WDEL,ODEL,MADEL,INTDEL,PER: TIME);
 port(DATA: inout WORD;
  MA: out ADDR;
  RUN,INT: in BIT;
  RD,WRITE,IO,RDS,RDP,WTS,WTP,INTA: out BIT);
end MARK2;
```

Figure 5.10 Entity definition for the MARK2.

Figure 5.11 shows an architectural body for the MARK2 entity, called BEHAVIOR. The first process in the body generates the internal clock for the processor. It is invoked when RUN rises, to generate the first clock, and then continues to generate clocks at interval PER. The only function of this clock is to trigger the STATE process whenever it makes either a positive or negative transition. Triggering the STATE process off either transition improves simulation efficiency, as there are no wasted clock events.

The RUN process is initiated by changes in signal RUN and either sets or resets the signal STOP. It also disables interrupts when RUN makes a 0-to-1 transition.

The process STATE is used to implement the state movement of the processor. There are three states: FETCH, EXECUTE, and INTERRUPT. Note that normal state changes take place in synchronization with the clock provided that RUN is stable, the processor is not stopped, and the processor is not in an I/O wait state.

The code within the processes for FETCH, EXECUTE, and INTERRUPT is activated whenever the corresponding states change to TRUE. Within each of these processes a check is made for a 0-to-1 transition using a wait statement at the beginning of the process.

The first activity in the FETCH process is to transfer the contents of the program counter (PC) to the memory address (MA) outputs. Next it sets up the RD and WRITE signals, and sets IOWAIT to 1. The wait command causes the process to suspend for RDEL nanoseconds. After the WAIT delay, the signals RD, WRITE, and IOWAIT are returned to their normal levels. The input data is then type converted and copied into the instruction register. The program counter is then incremented.

The WAIT period allows for I/O or memory response delay. During the WAIT period the signal IOWAIT is a logic 1 and this inhibits changes in processor state. Another variation on I/O waiting is possible (i.e., to wait for the I/O device to acknowledge the receipt of the I/O command). Thus if an acknowledge signal (ACK) were added to the entity signal list, one would replace the "wait for RDEL" with "wait on ACK until ACK = '1'."

In the EXECUTE process, the case statement is used to perform the instruction decoding function. "Wait" I/O is performed in three instructions and in the same manner as for the fetch process. It is left for the reader to verify that the VHDL code shown implements the instruction set.

The first activities in the INTERRUPT process are disabling interrupts (by setting INTE = 0), storing the return address, and setting INTA. INTA going to a logic 1 signals the interrupt controller to gate the interrupt vector address onto the data bus. After an I/O wait of duration INTDEL, the contents of the data bus (DATA) is copied into the program counter (PC). Note that a type conversion is required to do this. Since the next cycle is a FETCH cycle, the transfer to the interrupt routine location is achieved.

The IO_DECODE block generates the processor I/O control signals. When the guard for the block (BMA(4) and BMA(3)) is true, the guarded signal assignment statements are activated. Following the block is a single signal assignment statement which generates the control signal IO, which will be a logic 1 for I/O activities and logic 0 otherwise. The last group of statements in the body are conditional signal assignments used to perform time multiplexing of process outputs.

Some comments about the model structure. It is natural to use the wait construct for modeling the I/O interface. However, only a process can be made to wait. Thus the fetch, execute, and interrupt functions must be modeled as processes since they all perform I/O waiting. Thus the diagram in Figure 5.8 that shows the model structure, is a process model graph, as defined in Chapter 4. The block in the model, IO_DECODE, is a multiprocess node of the type referred to in the discussion of functional decomposition.

```
architecture   BEHAVIOR of   MARK2     is

      signal   STOP,STOPR,STOPE: BIT;
      signal   INTE,INTER,INTEE,INTEI: BIT;
      signal   BRD,RDF,RDE: BIT;
      signal   BWRITE,WRITEF,WRITEE: BIT;
      signal   BMA,MAF,MAE: ADDR;
      signal   IOWAIT,IOWAITF,IOWAITE,IOWAITI: BIT;
      signal   CLK,FETCH,EXECUTE,INTERRUPT: BOOLEAN;
      signal   PCF,PCI,PCE,PC: ADDR;
      signal   IR: BIT_VECTOR(7 downto 0);
      signal   INTRET: ADDR;
begin
 SYSCLK: process(CLK,RUN)
 begin
   if (RUN = '1') then
   CLK <= transport not CLK after PER;
   end if;
 end process SYSCLK;

 RUN_PROC: process(RUN)
 begin
   if RUN = '1' then
     STOPR <= '0';
     INTER  <= '0'; else
     STOPR <= '1';
   end if;
 end process RUN_PROC;
```

Figure 5.11   MARK2 architectural body.

```
STATE: process(RUN,STOP,IOWAIT,CLK,FETCH,EXECUTE,INTERRUPT)
begin
 if(not RUN'STABLE) and (RUN = '1') then
  FETCH <= true;
 elseif (RUN'STABLE and (STOP = '0') and (IOWAIT = '0')
              and not CLK'STABLE) then
  if ((EXECUTE and (INTE = '0' or INT = '0')) or INTERRUPT) then
   FETCH <= true; else
   FETCH <= false;
  end if;
  if FETCH then
   EXECUTE <= true; else
   EXECUTE <= false;
  end if;
  if (EXECUTE and (INT = '1' and INTE = '1')) then
   INTERRUPT <= true; else
   INTERRUPT <= false;
  end if;
 end if;
end process STATE;


FETCH_PROC: process
begin
 wait on FETCH until FETCH;
 MAF <= PC after MADEL;
 RDF <= '1'after ODEL; WRITEF <= '0'after ODEL;
 IOWAITF <= '1';
 Wait for RDEL;
 IR <= TRIVEC_TO_BITVEC(DATA);
 RDF <= '0'after ODEL;
 IOWAITF <= '0';
 PCF <= INC_ADDR(PC);
end process FETCH_PROC;


INTERRUPT_PROC: process
begin
 wait on INTERRUPT until INTERRUPT and INTE ='1';
 INTEI <= '0';
 INTRET <= PC;
 INTA <= '1'after ODEL;
 IOWAITI <= '1';
 wait for INTDEL;
 PCI<=VDAD(DATA);
 INTA <= '0'after ODEL;
 IOWAITI <= '0';
end process INTERRUPT_PROC;
```

Figure 5.11   MARK2 architectural body (continued)

```
EXECUTE_PROC: process
 variable ACC:  WORD;
begin
 wait on EXECUTE until EXECUTE;
 case(IR(7 downto 5) is
   when "000" => PCE  <= IR(4 downto 0);                    --jmp
   when "001" => ACC   := INC_WORD(INVW(ACC));              --tcm
   when "010" => MAE  <= IR(4 downto 0);                    --lda
    RDE <= '1'after ODEL; WRITEE <= '0'after ODEL;
    IOWAITE <= '1';
    wait for RDEL;
    RDE <= '0'after ODEL;
    IOWAITE <= '0';
    ACC := DATA;
   when "011" => DATA <= ACC after ODEL,                    --sta
                        "ZZZZZZZZ" after 3*ODEL + WDEL;
    MAE <= IR(4 downto 0) after MADEL;
    RDE <= '0'after ODEL;WRITEE <= '1'after ODEL;
    IOWAITE <= '1';
    wait for WDEL;
    WRITEE <= '0'after ODEL;
    IOWAITE <= '0';
   when "100" => MAE <= IR(4 downto 0);                     --add
    RDE <= '1'after ODEL; WRITEE <= '0'after ODEL;
    IOWAITE <= '1';
    wait for RDEL;
    RDE <= '0'after ODEL;
    IOWAITE <= '0';
    ACC := ADD8(ACC,DATA);
   when "101" =>                                            --int
    if IR(4) = '1' then
     INTEE <= '1';
    elsif IR(3) = '1' then
     INTEE <= '0';
    elsif IR(2) = '1' then
     PCE <= INTRET;
    end if;
   when  "110" =>
    if ACC(7) = '1' then                                    --jpn
    PCE <= IR(4 downto 0);
    end if;
   when  "111" => STOPE <= '1';                             --stop
 end case;
end process EXECUTE_PROC;
```

Figure 5.11   MARK2 architectural body. (continued)

```
-------------- I/O Control Signal Decoding
   IO_DECODE: block(BMA(4) and BMA(3) = '1')
   begin
RDP <= guarded not BMA(2) and not BMA(1) and not BMA(0) after ODEL,
                '0' after ODEL+RDEL;
WTP <= guarded not BMA(2) and not BMA(1) and BMA(0) after ODEL,
                '0' after ODEL+WDEL;
RDS <= guarded not BMA(2) and BMA(1) and not BMA(0) after ODEL,
                '0' after ODEL+RDEL;
WTS <= guarded not BMA(2) and BMA(1) and BMA(0) after ODEL,
                '0' after ODEL+WDEL;
   end block IO_DECODE;
   IO <= BMA(4) and BMA(3) after ODEL, '0'after ODEL + RDEL;

-------------- Process Output Multiplexing
   PC <= PCE when not PCE'quiet else
         PCI when not PCI'quiet else
         PCF when not PCF'quiet else PC;
   STOP <= STOPR when not STOPR'quiet else
           STOPE when not STOPE'quiet else STOP;
   IOWAIT <= IOWAITF when not IOWAITF'quiet else
             IOWAITE when not IOWAITE'quiet else
             IOWAITI when not IOWAITI'quiet else IOWAIT;
   INTE <= INTER when not INTER'quiet else
           INTEE when not INTEE'quiet else
           INTEI  when not INTEI'quiet else INTE;
   BMA  <= MAF when not MAF'quiet else
           MAE when not MAE'quiet else
           BMA;
   MA <= BMA;
   BRD <= RDF when not RDF'quiet else
          RDE when not RDE' quiet else
          BRD;
   RD <= BRD;
   BWRITE <= WRITEF when not WRITEF'quiet else
             WRITEE when not WRITEE'quiet else
             BWRITE;
   WRITE <= BWRITE;
end BEHAVIOR;
```

Figure 5.11 MARK2 architectural body. (continued)

RAM MODEL

In this section we present the RAM entity model. In fact, we give two architectural
bodies for the RAM. One model describes the basic behavior of the RAM, but represents the timing and logic structure crudely. A second model more accurately
portrays the logic and timing of a real RAM and thus fully meets our definition of
a chip-level model.

In Figure 5.12a the RAM entity is defined. Figure 5.12b gives the architectural body SIMPLE for the RAM. The body is implemented by a single process which is activated by a change in NCS (the low-active chip select line), RD (the READ line), and WRITE (the write line). Internal to the process a test for NCS low is made. If NCS is low, tests are made for positive transitions on RD and WRITE, and if one of them occurs, the operation indicated is performed. If NCS is high or if RD goes low while NCS is low, the high-impedance condition "ZZZZZZZZ" is tranferred to the bus DATA. Note that since all process variables are static, the variable MEM will preserve memory contents. Also a type conversion is employed in memory accessing; that is, type ADDR, an array of bits, is converted to type INTEGER to index the array representing the RAM.

```
entity  RAM is
  generic(RDEL,DISDEL: TIME);
  port
  (DATA: inout WORD;
   ADDR: in ADDR;
   RD,WRITE,NCS: in  BIT);
end  RAM;
```

Figure 5.12(a)   The RAM entity.

```
architecture SIMPLE of RAM         is
begin
  process  (NCS,RD,WRITE)
  type MEMORY is array(0 to 31) of WORD;
  variable MEM: MEMORY;
  begin
  if  NCS = '0'   then
   if not RD'STABLE then
    if RD = '1' then
     DATA <= MEM(INTVAL(ADDR)) after RDEL;
    else
     DATA <= "ZZZZZZZZ" after DISDEL;
    end if;
   elsif WRITE = '1'and not WRITE'STABLE   then
       MEM (INTVAL(ADDR)) := DATA;
   end if;
  else
   DATA <= "ZZZZZZZZ"   after DISDEL;
  end if;
  end process;
end SIMPLE;
```

Figure 5.12(b)   Ram architectural body SIMPLE.

The simple model would be very useful for initial system architectural verification but is inaccurate in a number of areas. First, in terms of logical operation, the model is deficient because only changes in NCS, RD, or WRITE invoke the process that implements the model. In fact, for a real RAM, if the RAM is selected (NCS = 0), and if either READ or WRITE is high, changes in the input address will affect the memory. Normal operation may prescribe that the address be stable before the select signal goes low, but this is a system-level constraint. Designing individual models to meet system-level constraints limits their generality.

A second deficiency is in the area of timing modeling. Architecture SIMPLE models the read delay (RDEL) and disable delay (DISDEL) only. No write delay is modeled, nor is there any modeling of the necessary time relationship between the RAM inputs. For example, typical RAM input timing specifications are the following:

READ:

    Access Time (Tacc)   350 ns

    Chip Enable
    to Output Time (Tco)   180 ns

WRITE:

    Write Pulse Width (Twp)   250 ns

    Data Setup Time   (Tdw)   250 ns

    Chip Enable to
    Write Set-Up Time (Tcw)   250 ns

Figure 5.13 illustrates the timing for these signals. However, to apply these signals intelligently, some knowledge of internal memory structure is required. Figure 5.14 shows the internal structure of a RAM which could have the specifications given above. Note that the address is decoded to generate ROW SELECT and COLUMN SELECT signals. The ROW SELECT signal is applied directly to the storage cell, but the COLUMN SELECT is ANDed with the CHIP ENABLE, WRITE, and DATA IN to activate a write operation and is ANDed with the CHIP ENABLE and READ to activate a read operation.

Consider the read operation first. Since the access time (Tacc) is 350 ns and the chip enable to output time (Tco) is 180 ns, one can infer from Figure 5.14 that the decoder delay is 170 ns. As far as write is concerned, Twp, Tdw, and Tcw all have the same value (250 ns) since, for write, signals WRITE, DATA IN, and CHIP ENABLE all propagate through the same data path.

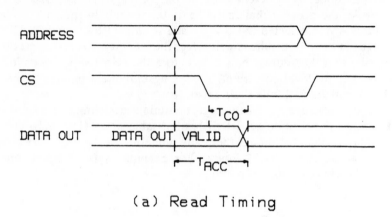

(a) Read Timing

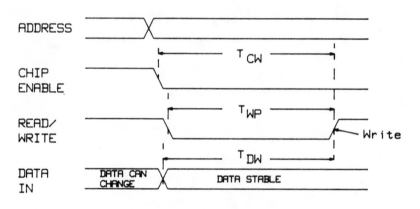

(b) Write Timing

Figure 5.13   RAM timing.

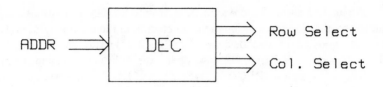

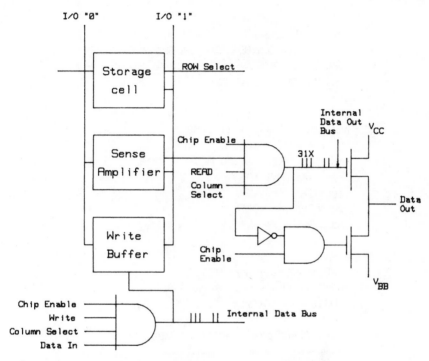

Figure 5.14   Internal data paths for RAM cell.

Figure 5.15 shows a more realistic RAM model. The major timing delays are modeled by the first three assignment statements in the architectural body. Three delays are modeled: (1) decode delay (DECDEL), (2) write select delay (WSDEL), and (3) data delay (DDEL). Because of the inertial delay mechanism built into the VHDL "after" clause, DECDEL, WSDEL, and DDEL also serve as setup times; that is, if the signal does not persist for that period of time, no signal change will be propagated.

```
entity RAM2 is
  generic(DECDEL,WSDEL,DDEL,RDEL,DISDEL: TIME);
  port(READ,WRITE,CS: in BIT;ADDR: in ADDR;
       DATA: inout WORD);
end RAM2;                                    .


architecture REALISTIC of RAM2 is

  type MEMORY is array(0 to 31) of WORD;
  signal MEM: MEMORY;
  signal DDATA: WORD;
  signal DADDR: ADDR;
  signal WSEL: BIT;

begin
  DADDR <= ADDR after DECDEL;
  WSEL <= WRITE and CS after WSDEL;
  DDATA <= DATA after DDEL;

  READ_PROC: process(DADDR,CS,READ)
  begin
   if CS = '1' then
    if READ = '1' then
     DATA <= MEM(INTVAL(DADDR)) after RDEL;
    end if;
   else
    if not CS'STABLE then
    DO <= "ZZZZZZZZ" after DISDEL;
    end if;
   end if;
  end process READ_PROC;
```

Figure 5.15   Realistic RAM model.

```
WRITE_PROC: process(WSEL,DDATA,DADDR)
begin
  if WSEL = '1' then
    MEM(INTVAL(DADDR)) <= DDATA;
  end if;
end process WRITE_PROC;

end REALISTIC;
```

Figure 5.15   Realistic RAM model. (continued)

The READ process is activated by changes in DADR, CS, and READ. If chip select (CS) is logic 1, a read operation is performed. The delayed value of the address (DADDR) is used for this read; the output is changed after a delay of RDEL nanoseconds. If CS is not a logic 1, the output is disabled after DISDEL nanoseconds. That the disable delay is different and shorter than the read delay can be seen by examining Figure 5.14. The WRITE process is activated by a change in WSEL, DDATA, or DADDR. If WSEL is a logic 1, writing takes place using DDATA and DADDR.

The five delays in the model are declared as generics in the entity description. To implement the specifications given above, they would be given the following values when the component was instantiated:

DECDEL = 170 ns   WSDEL = 250 ns   DDEL = 250 ns

RDEL = 180 ns   DISDEL = 100 ns*

The last value, DISDEL, is starred because the value cannot be inferred directly from the specification, and is an estimate.

UART MODEL

Figure 5.16 gives the block diagram of the UART used in the system and Figure 5.17 shows its VHDL description. The rising edge of the control input LOAD loads the DATA input values into the output register OREG and sets the low-active output interrupt signal (NINTO) high. After that, the contents of OREG are transmitted to the output (O) serially, starting from the highest bit. The duration of the serial bit time is controlled by the internal output shift clock (OCLK). When the output shifting process is complete, the interrupt output NINTO is set low.

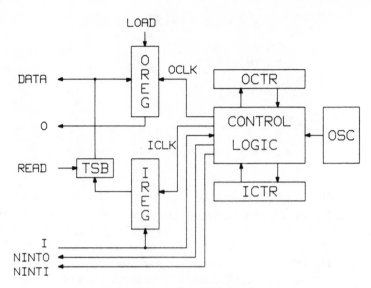

Figure 5.16 UART block diagram.

Receiving of the serial input (I) is initiated by the first negative transition of I (starting bit). The logic then delays for half a clock period (CLK_PER/2) and then samples I again. If I is still a logic 0, the sampling process continues. Note that the logic is now sampling in the middle of the input clock period, which is the most reliable place to sample. The input I is sampled every CLK_PER nanoseconds until eight samples are received. These values are stored in the input register IREG. When the input shifting process is complete, the interrupt output (NINTI) is set low.

The processor responds to the NINTI interrupt from the UART by performing a READ operation. When the READ control input rises, the contents of IREG are transferred to the DATA output, and the NINTI output is set to 1.

```
entity UART is
  generic(CLK_PER,ODEL,INDEL,INTDEL: TIME);
  port(DATA: inout WORD;
       I: in TSL;
       LOAD,READ: in BIT;
       O: out TSL;
       NINTO,NINTI: out BIT);
end UART;
```

Figure 5.17 UART Model.

```
architecture     BEHAVIOR of UART        is

signal ICLK,OCLK,ISTRT: BIT := '0';
signal NINTI1,NINTI2,TEMP_NINTI: BIT := '1';
signal IREG: WORD;

begin
OUTPUT: process (LOAD,OCLK)
  variable OCNTR: INTEGER;
  variable OREG : WORD;
begin
  if not LOAD'STABLE   and LOAD = '1'   then
  OREG := DATA;
  OCNTR := 7;
  NINTO <= '1' after INTDEL;
  O <= '0'after ODEL;
  OCLK <= not OCLK after CLK_PER;
  end if;

  if not OCLK'STABLE      then
  if OCNTR /= -1   then
    O <= OREG(OCNTR) after ODEL;
    OCNTR := OCNTR-1;
    OCLK <= not OCLK after CLK_PER;
  else
    O <= '1'after ODEL;
    NINTO <= '0' after INTDEL;
  end if;
  end if;

end process OUTPUT;

INP: process(I,ISTRT,ICLK)

  variable ICNTR: INTEGER;
  variable I_FLAG: BOOLEAN := TRUE;

begin
  if I_FLAG then
  if not I'STABLE and I= '0'   then
    ISTRT <= '1' after CLK_PER/2;
  end if;
  end if;
```

Figure 5.17   UART Model. (continued)

```
if not ISTRT'STABLE and ISTRT = '1' and I = '0' then
  I_FLAG := FALSE;
  ISTRT <= '0';
  ICNTR := 7;
  IREG(ICNTR) <= I;
  ICLK <= not ICLK after CLK_PER;
end if;

if not ICLK'STABLE then
  if ICNTR /= -1  then
    IREG(ICNTR) <= I;
    ICNTR := ICNTR-1;
    ICLK <= not ICLK after CLK_PER;
  else
    NINTI1 <= '0' after INTDEL;
    I_FLAG := TRUE;
  end if;
end if;
end process INP;

READ_PROC: process(READ)
begin
  if READ = '1'  then
    DATA <= IREG after INDEL;
    NINTI2 <= '1' after INTDEL
  else
    DATA <= "ZZZZZZZZ" after INDEL;
  end if;
end process READ_PROC;

TEMP_NINTI <= NINTI1 when not NINTI1'quiet else
              NINTI2 when not NINTI2'quiet else
              TEMP_NINTI;
NINTI <= TEMP_NINTI;
end BEHAVIOR;
```

Figure 5.17   UART model. (continued)

PARALLEL I/O PORTS

For the input and output parallel ports in the system, we will employ the Intel 8212 buffered latch. A logic diagram of the latch is given in Figure 5.18. The I8212 has control inputs DS1, DS2, MD and STB. These inputs are used to

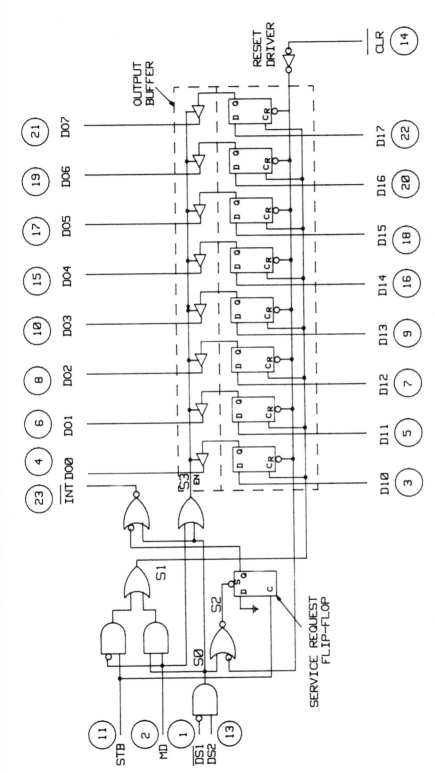

Figure 5.18 Intel 8212 buffered latch.

control device selection, data latching, output buffer state, and the service request flip-flop. When DS1 is low and DS2 is high, the device is selected. When MD is high (output mode), the output buffers are enabled and the source of clock to the data latch is the device selection logic. When MD = 0 (input mode), the STB input is used as the clock to the data latch, and to reset synchronously the service request flip-flop (SRQ). SRQ is negative edge triggered. SRQ is set when CLR is low or the device is selected. Also when MD = 0, the output buffers are enabled whenever the device is selected.

Figure 5.19 is the VHDL description which is partitioned into two sections: the latch section and control section. The latch section is a guarded block that implements a D latch with overriding clear. It also contains the tristate buffer logic. Part of the control section is implemented as data flow logic. While other higher-level implementations are possible, this is the most straightforward approach if one wishes to incorporate realistic time delays. The model timing uses gate (GDEL), flip-flop (FFDEL), and buffer (BUFDEL) delays. Also, the amount of modeled logic is small, so that use of the process block would involve unnecessary overhead.

```
entity I8212 is
generic(GDEL,FFDEL,BUFDEL: TIME);
port(DI: in   WORD;
     DO: out WORD;
     NDS1,DS2,MD,STB,NCLR: in BIT;
     NINT: out BIT);
end I8212;

architecture BEHAVIOR   of   I8212      is

signal S0,S1,S2,S3: BIT;
signal SRQ: BIT;
signal Q,Q1,Q2: WORD;

begin
block (S1 = '1'and NCLR = '1')
begin
  Q1 <= guarded DI after FFDEL;
  Q2 <= "00000000" after FFDEL   when   (NCLR='0') else Q2;
  DO <= Q after BUFDEL when (S3='1')   else
        "ZZZZZZZZ" after BUFDEL;
end block;
```

Figure 5.19   Parallel port model.

```
S0 <= not NDS1 and DS2 after GDEL;
S1 <= (S0 and MD) or (STB and not MD) after (2*GDEL);
S2 <= S0 nor not NCLR after GDEL;
S3 <= S0 or MD after GDEL;

SERVRQ: process( S2, STRB)
begin
 if (S2 = '0') then
   SRQ <= '1' after FFDEL;
 elsif (S2 = '1') and (not STRB'STABLE) and (STRB = '0') then
   SRQ <= '0' after FFDEL;
 end if;
end process SERVRQ:

NINT <= not SRQ nor S0 after GDEL;
Q <= Q1 when not Q1'QUIET else
     Q2 when not Q2'QUIET else
     Q;
end BEHAVIOR;
```

Figure 5.19   Parallel port model. (continued)

INTERRUPT CONTROLLER

Figure 5.20 shows the interrupt controller for the system. Four active-low interrupt request lines are provided: NINT1, NINT2, NINT3, and NINT4. NINT1 is the highest priority. The four NINT lines are NANDed together to generate the interrupt request

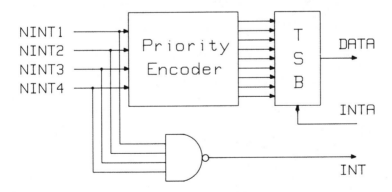

Figure 5.20   Interrupt controller.

signal to the processor (INT). The priority encoder converts the highest-priority active interrupt into an interrupt vector. The interrupt vector will be gated through the tristate buffers onto the bus when INTA goes to a 1. It is assumed that a given interrupt request is cleared by the processor at the beginning of the interrupt service routine. Also, the processor will not respond to other interrupt requests until the interrupt service routine is finished.

Figure 5.21 gives the VHDL description for the interrupt controller. The architectural body consists of three processes. The first single statement process generates the interrupt request (INT). Process VECTOR performs the encoding function. Process BUFFER implements the tristate buffer function.

```
entity INT_CONT
  generic(INTDEL,ENCDEL,BUFDEL: TIME);
  port(NINT1,NINT2,NINT3,NINT4,INTA: in BIT;
                INT: out BIT ; VOUT: out WORD);
end INT_CONT;

architecture BEHAVIOR of INT_CONT is

signal VECT: WORD: = "00011100";

begin
INT <= not(NINT1 and NINT2 and NINT3 and NINT4) after INTDEL;

VECTOR: process(NINT1,NINT2,NINT3,NINT4)
begin
 if (NINT1 = '0') then
  VECT <= "00011100" after ENCDEL;
 elsif  (NINT2 = '0') then
 VECT <= "00011101" after ENCDEL;
 elsif  (NINT3 = '0') then
 VECT <= "00011110" after ENCDEL;
 elsif  (NINT4 = '0') then
 VECT <= "00011111" after ENCDEL;
 else
 VECT <= "00011100" after ENCDEL;
 end if;
end process VECTOR;

BUFF: process(INTA)
begin
 if (INTA = '1') then
  VOUT <= VECT after BUFDEL;
 else
  VOUT <= "ZZZZZZZZ" after BUFDEL;
 end if;
 end process BUFF;
end BEHAVIOR;
```

Figure 5.21

SUMMARY

In this chapter we have discussed two main issues: modeling of system interconnection and modeling of generic logic structures for systems. In discussing the modeling of system interconnection we first presented a model based on the use of signals. Next, we showed how interconnection could be modeled at varying levels of abstraction using the data typing capability of VHDL. Finally, the concepts of bused signals and time multiplexing were explained.

A complete system example illustrated generic logic structures. The system example included the complete definition of the packages required for the system model.

BIBLIOGRAPHY

Short, K., *Microprocessors and Programmmed Logic*, 2nd ed., Englewood Cliffs, N.J., Prentice-Hall, Inc., 1980.

Sieworek, D., C. Bell, and A. Newell, *Computer Structures: Principles and Examples.* New York: McGraw-Hill Book Company, 1982.

6

OTHER MODELING
ISSUES

In the previous chapters, we have addressed one aspect of chip-level modeling: modeling the behavior of a good device using a hardware description language. However, there are other major issues related to chip-level modeling and the use of hardware description languages. In this chapter we discuss: chip-level fault modeling and test generation, chip-level simulation in parallel, and aids to the chip-level modeling process. Other problems related to chip-level modeling and the use of hardware description languages are also outlined.

CHIP-LEVEL FAULT MODELING AND TEST GENERATION

Since one major application of digital system modeling is test generation, it is important to be able to model faulty behavior effectively. In this section we discuss chip-level fault models, an approach to chip-level test generation, and methods for measuring chip-level fault coverage.

Basic Requirements

One of the major challenges for chip-level modeling is the development of an effective fault modeling scheme. Previous efforts at the chip level, or other forms of functional fault modeling, have been "ad hoc" in that faults were selected by the modeler somewhat arbitrarily. With this approach, one has no concept of a complete fault set and thus has no well-understood measure of coverage. To address these concerns, any proposed method of fault modeling must have:

1. A well defined fault model
2. An algorithmic method for fault selection and test generation
3. An accepted measure of test vector fault coverage

As an example of the application of these criterion, consider what is done at the gate level. At this level, the fault model is the "stuck-at" fault model. Fault selection and test generation is generally achieved through some combination of the D algorithm and register scan techniques, and coverage is defined as the percentage of single, stuck-at faults covered. Below we show how the three requirements listed above can be met using chip-level modeling. Following the theme of the book, the models are expressed in VHDL.

Chip-Level Fault Model

In chip-level modeling, a model of a VLSI device is constructed as a single logical entity. The basic structure of a chip-level model is that of a sequence of micro-operations. This structure is illustrated in Figure 6.1, which shows a model procedure in which a micro-operation sequence leads to a control point where branching occurs to one of two other micro-operation sequences. Figure 6.2 shows the proposed fault model for chip-level faults. One class of faults involves faulting the individual micro-operations. We shall designate these faults as FMOP faults. A second class of faults, which involves faulting the control points that switch between sequences, will be referred to as FCON faults.

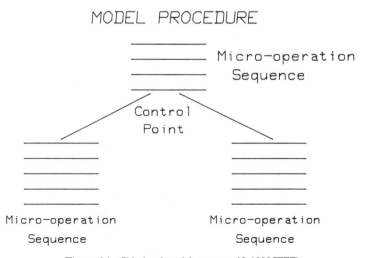

Figure 6.1 Chip-level model structure. (© 1988 IEEE).

FAULT MODEL

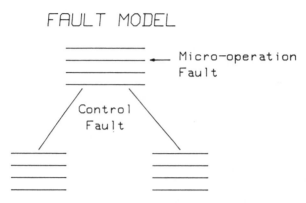

Figure 6.2   Chip-level fault model (© 1988 IEEE).

We discuss control faults first. Control is achieved by means of such constructs as IF-THEN-ELSE or CASE. The IF construct can be faulted in (at least) two basic ways. First, it can be assumed to be stuck such that branching will always occur in one direction independent of control signal values (i.e., so-called "stuckthen" and "stuckelse" faults). The test generation algorithm described below will use this approach. A second approach is to invert the sense of the IF statement, that is, assume that IF(X) is transformed to IF(not X). CASE statements can be faulted by assuming that a clause selected by the CASE statement does not execute. Consider the following example:

```
case X(0 to 1) is
    when "00" => A <= B;
    when "01" => A <= not B;
    when "10" => A <= '1';
    when "11" => A <= '0';
end case;
```

There are four clauses in this case statement; thus four "dead clause" faults would be modeled.

With micro-operation faults, it is assumed that the effect of the fault is to perturb one micro-operation to another. A difficult question to answer is: Which micro-operation should we perturb to? The question is difficult to answer since there are so many possible choices. What is required is a heuristic rule to aid in this choice. One such rule is that of replacing a micro-operation by its logical dual. For example, the micro-operation Z <= X and Y could be perturbed to Z

<= X or Y. Two comments about this approach. First, note that the logical dual substitution takes place at the level of the modeling constructs built into the hardware description language (e.g., and, or, xor, etc.). It is not applied to complex boolean functions. Second, the logical dual is defined only for boolean functions, so in the case of arithmetic operations, one has to choose a dual (e.g., xor for add).

Another question that one might ask is: How does one know whether any of these micro-operation faults can actually occur? The answer is that one has no real "a priori" knowledge whether a particular micro-operation fault is likely to occur. However, such faults represent basic model perturbations, and test vectors which are designed to detect a complete set of such faults would thoroughly exercise the logic of the chip. When we discuss fault coverage below, we assess the effectiveness of this approach.

Chip-Level Test Generation

For fault modeling to be useful, the process of fault selection and test vector development must be amenable to automation. What is required is a computer program which, when given a chip-level description of a VLSI chip in some hardware description language, will systematically select faults and determine a test vector set for the faults. The program would continue to do this until the complete set of chip-level faults had been selected and test vectors developed for them. Completeness would be defined in terms of the structure of the HDL description (i.e., the number of micro-operations and control points in it).

We discuss next a promising approach to the solution of this problem developed by Daniel Barclay. The treatment here is necessarily brief. Details of the approach are given in the bibliography at the end of the chapter.

The approach employs the FCON and FMOP fault models discussed above and attempts to develop an equivalent of the D algorithm for HDL descriptions of the chip logic. The approach works as follows. Given a faulty micro-operation (FMOP) or control operation (FCON), a test is developed by (1) activating the faulted operation (fault sensitization), (2) propagating the effect of the fault through other statements to a statement that will transfer the effect to an output (path sensitization), and (3) determining input combinations that will justify the path sensitization. The algorithm can perform this process for descriptions that represent sequential as well as combinational circuits. The models are represented in VHDL, and the form of the representation is restricted to data flow models. This is because the steps given above are too difficult to perform for algorithmic models containing such features as loops and wait statements.

In developing the algorithm, Barclay used the artificial intelligence concept of goal trees to organize the algorithm structure. With this approach, each node in the tree represents a step or goal in the development of a test vector. The process of developing and simplifying the goal tree involves breaking goals down into subgoals. The process is complete when all goals are broken down into "primitive goals" such as specifying circuit input conditions. Table 6.1 shows the goal types for the test generation algorithm.

Table 6.1   Goal Types

VIO	EXEC	OBSOBJ	DND
VIE	DNE	OBSEXPR	TR
	EXG	OBSEXEC	

Since the algorithm is designed to develop tests for sequential logic, a test, in general, will consist of a sequence of test vectors, and all the goals in the table have a time mark associated with them (i.e., the goal must be satisfied at a certain time). Time is measured at discrete intervals (e.g., t_0, $t_0 + 1$, $t_0 - 1$). The goals in Table 6.1 have the following definitions:

1. VIO(VIE). Need a value in an object (expression) at time T.
2. EXEC(DNE). Execute (don't execute) a statement at time T.
3. EXG. Execute a statement at time T given that parent statements execute.
4. OBSOBJ(OBSEXPR). Observe a value in an object (expression) at time T for good or bad value.
5. OBSEXEC. Observe execution of clauses of statement at time T, expecting good or bad clauses to execute.
6. DND. Do not disturb the value in an object for a given time period.
7. TR. Specify a relationship between two times (e.g., $t_1 = t_2$, or t_1 greater than or equal to t_2).

The term *object* used in the definitions refers to a signal in the VHDL description.

We now illustrate the use of this goal tree structure with a simple example. Figure 6.3 gives the logic diagram of circuit consisting of a single D flip-flop and an AND gate. The output of the circuit is the logical AND of the input to the D flip-flop and its output. From the testing point of view, the signals D, CLK, and NCLR are test inputs, and signal OUT is a test output.

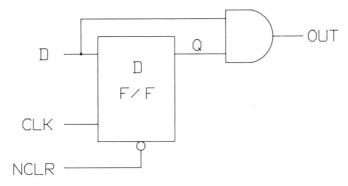

Figure 6.3   D flip - flop circuit.

Figure 6.4 shows the VHDL description for the test generation example. The VHDL is standard except that statement labels have been added for reference purposes. As far as model characteristics are concerned, note that the D flip-flop is implemented as being positive edged triggered with an overriding clear.

```
entity  D_AND
  port(NCLR,CLK  in  BIT;  OUT  out  BIT);
end  D_AND;

architecture  DATA_FLOW  of  D_AND  is
  singal  Q:  BIT;
begin
 process(NCLR,CLK,Q)
  begin
   S1:  if  NCLR  =  '0'  then
   S2:    Q  <=  '0';  else
   S3:    if  (CLK  =  '1'and  not  CLK'STABLE)  then
   S4:    Q  <=  D;
          end  if;
        end  if;
   S5:  OUT  <=  Q  and  D;
  end  process
end  DATA_FLOW
```

Figure 6.4   VHDL description of test generation example.

Barclay's algorithm first generates the fault list shown in Figure 6.5. The list gives chip-level faults for each statement; for example, statement S3 has control faults stuckthen and stuckelse and micro-operation faults for the AND, equivalence, and NOT operators in the expression tested by the if statement. Statements S2, S4, and S5 have "assignment" faults associated with them. This fault assumes that the assignment fails to work; that is, the target of the assignment remains unchanged.

The fault list is important because it allows us to define complete fault coverage in the chip-level sense.

STATEMENT	FAULTS
S1	stuckthen,stuckelse,microp(=)
S2	assignment
S3	stuckthen,stuckelse,microp(and)microp(=),microp(not)
S4	assignment
S5	assignment,microp(and)

Figure 6.5   Fault List.

Once the fault list is complete, the test generation process can begin. We illustrate this process for the fault: microp(AND) for S5. The micro-operation fault modeling assumes that the AND micro-operation is perturbed to an OR. A test for this perturbation is that the two inputs to the micro-operation, Q and D is this case, are different. Thus a test for the fault would be to set $Q = 1$ and $D = 0$ and observe the output of the expression "Q and D," which will be 1 in the presence of the fault and 0 otherwise. The algorithm determines this and then sets up three goals, which must all be solved:

G1: OBSERVEXP(Q and D) at time t_0
for good(0) or bad(1) value;
G2: VIO at time t_0: $Q = 1$;
G3: VIO at time t_0: $D = 0$;

Assume that it tries to solve G1 first. The algorithm quickly determines that this expression is copied into the object OUT. Thus since statement S5 always executes, goal G1 is equivalent to the goal G4: OBSERVOBJ(OUT) at time t_0 for good(0) or bad (1) value; but this goal is primitive (i.e., inherently solved), as OUT is an observable test output. G3 is also primitive since D is a controllable test input. The remaining goal G2 is more complicated. There are two statements (S2 and S4) which load Q. Thus to solve G2, one has two options. Either solve both the goals

G5: EXECUTE S2 at time t_0
G6: VIE at time t_0-1: $0 = 1$

or both the goals

G7:   EXECUTE   S4 at time t_0

G8:   VIE at time t_0 - 1: D = 1

Assume that the option (G5, G6) is taken first. Goal G6 is unsolvable and thus back-tracking must occur. This causes the algorithm to try the other option (G7, G8). G8 is equivalent to the goal G9: VIO at time t_0-1: D = 1, which is primitive. G7 requires that the following subgoals be solved:

G10: EXECUTE S3 at time t_0;

G11: VIE at time t_0: (CLK = '1'and not CLK'STABLE) = TRUE;

In turn, G11 requires that the following goals be solved:

G12: VIO at time t_0: CLK = 1;

G13: VIO at time t_0 - 1: CLK = 0;

As CLK is a test input, these two goals are primitive. To solve G10 requires that

G14: VIO at time t_0: NCLR = 1;

G15: EXECUTE S1 at time t_0;

be solved. Both of these goals are primitive since NCLR is a test input and S1, being the outermost if statement, always executes. The test generation process is complete as the original set of three goals (G1, G2, G3) is now solved.

Figure 6.6 shows the goal tree for the fault generation process that we have just described. The test for the fault is given by the following set of primitive goals:

G3: VIO at time t_0: D = 0;

G4: OBSERVOBJ(OUT) at time t_0 for good(0) or bad(1) value;

G9: VIO at time t_0-1: D = 1:

G12: VIO at time t_0: CLK = 1;

G13: VIO at time t_0 - 1: CLK = 0;

G14: VIO at time t_0: NCLR = 1;

G15: EXECUTE S1 at time t_0;

Thus the algorithm generates the test shown in Figure 6.7.

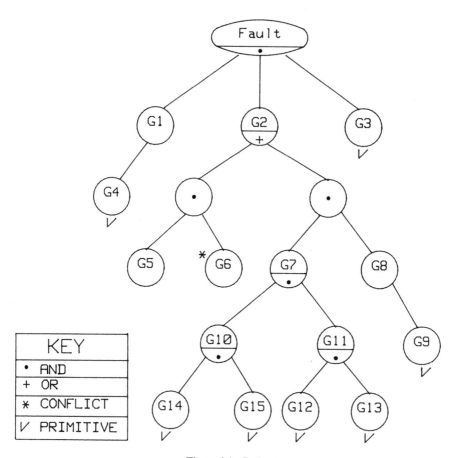

Figure 6.6 Goal tree.

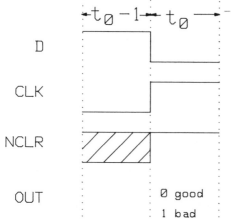

Figure 6.7 Generated test.

The test generation example given here is necessarily simple. For more complicated examples the reader can consult Barclay's master's thesis, which is listed in the bibliography at the end of the chapter.

Barclay implemented his algorithm in Prolog. Left to its own devices, Prolog will solve goals in lexical order. The solving process involves searching, which is computationally very intensive. Efficiency can be improved by choosing the order in which to solve goals by using heuristics based on observability and controllability considerations. Partitioning the problem so that different parts of a goal tree are searched by different parallel processors can also speed up the test generation process.

Fault Coverage for Chip-Level Modeling

In the example given above the test algorithm would generate a test for each chip-level fault. This set of tests would then cover all the chip-level faults (i.e., the fault coverage would be 100%). However, at the present time it is doubtful that chip-level fault coverage would be accepted in the computer industry as a measure of test vector quality, as the gate-level fault model is a de facto standard. To address this problem, one can relate the chip-level fault coverage of a test vector set to its gate-level fault coverage by conducting the following experiment:

1. Develop a chip-level model for the given VLSI device.
2. Determine a set of chip-level faults using the FCON and FMOP fault models.
3. Develop a set of test vectors that will detect the complete chip-level fault set.
4. Apply the set of test vectors developed at the chip level to a gate-level model of the logic structure, and use traditional fault grading to determine what percentage of stuck-at faults are covered by the test vector set.

In experiments conducted at the Electrical Engineering Department at Virginia Tech, this approach was used on a set of 11 logic circuits representing a cross section of generic types of logic. The average equivalent gate-level coverage for the experiments on these circuits was 92%. Analysis of the results indicated that the chip-level tests did not cover data path faults well. The coverage could be improved significantly (e.g., to 98%) by adding heuristics that generate "variety" for data values for which the basic chip-level method chooses arbitrary values. The manner in which these heuristics would work in conjunction with the basic test generator is shown in Figure 6.8. Note that the standard test generator generates constraints on input signal values which solve the required goals. The heuristic test generator responds by providing a test vector or set of vectors which meets these constraints. As an example, suppose that the standard test generator produces the following constraints on a signal A (i.e., A = 01XX). Then the heuristic test generator fills in the don't-care positions, perhaps based on the previous history of the bit patterns applied to A (e.g., by trying combinations not yet used).

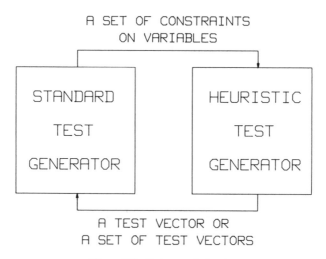

Figure 6.8 Test generation system.

Defining fault coverage for chip-level modeling in terms of gate-level coverage is an empirical approach, but it does relate the effectiveness of test vectors developed at the chip level to a measure with which the industry is familiar. It can be used to build confidence in the chip-level approach and can ultimately lead to acceptance of chip-level fault coverage as a reliable measure of test vector effectiveness. It should also be emphasized that vectors developed from a chip-level model will detect some faults that gate-level models cannot even represent. This is especially true for VLSI circuit technology. In fact, experimental results reported by Shen, Maly, and Ferguson from Carnegie–Mellon University show that when random defects are injected into a CMOS VLSI layout, at most 64% of the defects result in circuit behavior that can be modeled as a stuck-at fault.

CHIP-LEVEL SIMULATION IN PARALLEL

One of the major reasons for developing digital system models is to simulate the behavior of the system in response to input stimuli. Thus simulation efficiency is an important requirement. In Chapter 1 we discussed simulation efficiency and stated that the nature of the chip-level model should make it more efficient to simulate than a gate-level model of the same device. However, CAD researchers are aiming to develop a capability whereby simulation models of complicated VLSI systems would be used for extensive evalution of good and faulty system performance. Obviously, to perform simulations of this type, parallel approaches to simulation are required.

For over 20 years gate-level simulators have employed bit parallelism for fault simulations running on a single processor. In the 1980s, simulation engines have been developed which have sped up gate-level simulation by almost a factor of 1000. Similar developments in computer architecture for chip-level simulation are required.

Two attributes of logic simulation make it an attractive candidate for parallel processing. First, the computational problem can be partitioned quite easily using the structure of the circuit being simulated. Second, once a simulation process has begun, no input/output is required to maintain the simulation. This assumes that no swapping is done to secondary memory and that no intermediate results are required. Chip-level modeling fits very naturally into this schema. The chip-level models provide the natural partitioning for the parallel approach.

A parallel approach to chip-level simulation was developed by Manos Roumeliotis in his doctoral dissertation, and the subsequent discussion follows his work. Figure 6.9 shows a proposed architecture for a system that performs chip-level simulation in parallel. Four processors are shown, each containing two chip-level models. There are two types of signal flow between models: intraprocessor flow in which signals between simulation models remain within a processor, and interprocessor flow, where signals must travel between processors using a high-speed interconnection network. The time queue in the parallel system is distributed, with each processor maintaining a local queue for events intended for it. Thus if model 1 in processor 1 generates a signal for model 7 in processor 4, processor 1 must use the interconnection network to deposit the signal event for model 7 in processor 4's time queue.

An important problem in distributed simulation is how the global simulation time is kept; that is, care has to be taken so that a processor does not execute too far ahead of the other processors and produce output signal events at an improper time. This is accomplished by a synchronization mechanism in which all processors read the current simulation time of other processors. The system time (Ts) is then computed by all processors as the minimum of all these times. Once each processor knows the system time, a number of approaches are possible.

In the simplest approach, a processor can execute if (1) its next event time equals Ts or (2) its next event time is greater than Ts but it receives signals from other processors at Ts. In this approach, processors with only local events cannot proceed with execution if their next event time is greater than Ts. A second approach avoids this difficulty by allowing these waiting processors to predict if they expect to receive any signals before their next local event time. If they expect no signals, they can go ahead and simulate locally. This prediction is carried out as follows. Processors know from which other processors they expect to receive signals. This information is stored in a system connectivity table, of which each processor has a copy. Also, each processor knows the next event time for every other processor. Thus processor i can predict whether any of its "drivers" will send it signals before its next event time. If the prediction indicates that it will receive no signals, processor i can proceed with simulation.

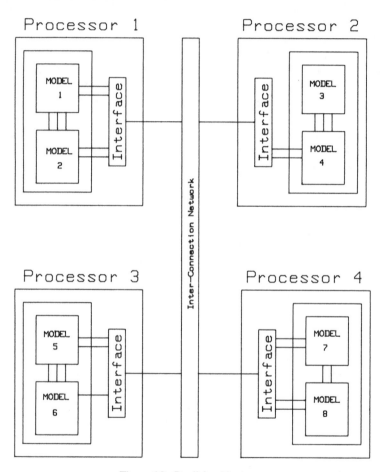

Figure 6.9   Parallel architecture.

There is one way this method can be foiled, however. After the time the prediction is made and processor i proceeds with simulation, another system processor, which is not a member of the drivers of i, sends a signal to one of the drivers of i. This processor could then send a signal to i at a simulation time which is earlier than i's next event time. To account for this possibility, each processor, when it computes ahead of the current simulation time, must preserve the previous state of simulation so that it can roll back to account for this unexpected input.

Roumeliotis analyzed this system configuration and showed that it could perform chip-level simulation with a speedup of $0.5N$, where N is the number of processors in the system. This is for the case of simulating a fault-free system where the models of the system are partitioned among the processors as in Figure 6.9. The speedup is related to the partitioning, as intraprocessor events are processed more efficiently than interprocessor events. Graph-theoretic techiques can be used to do this partitioning in an optimum manner. The basic idea is to cluster the models so that most signal activity is within a cluster, and then assign clusters to processors. He also showed how the system could perform fault simulation with a speedup of $0.8N$. Here there are C copies of a given system to be simulated simultaneously, one copy of the good system and $C - 1$ copies of the systems with different faults in them. The speedup is higher in this case since the communication requirements among all processors are less. A third potential application of the parallel processing architecture is simulating C copies of a system, all of which are good. This would be done for validation of system designs (i.e., each copy of the system would receive a different set of test vectors). The interprocessor communication requirements for this application would be lower than the fault modeling case and thus a speedup in excess of $0.8N$ could be expected.

Roumeliotis' work shows promise for speeding up chip-level simulation. What is required now is a commercial implementation of the concept.

ASSISTS TO THE CHIP–LEVEL MODELING PROCESS

One of the difficulties with chip-level modeling is that the modeling process involves a significant programming effort. Each model is a behavioral representation of a block of digital logic which must be coded in a hardware description language (e.g., VHDL). Also, model style will vary, depending on the approach taken by the coder. Thus models developed by one person may be difficult to understand by another. What is required is an approach to model development that is structured and relieves the modeler of many of the details of manual code entry. We shall propose that these features be incorporated into what we shall term the "Modeler's Assistant."

The general problem of automated program development is perhaps unsolvable, but we are not trying to solve the general case. The logic that is being modeled has many recurrent features; thus the representation for those features can be precoded. Also, many approaches to modeling are used repeatedly.

What is needed is a structure that will allow the modeler to take advantage of this situation. Such a structure might be provided by the process model graph introduced in Chapter 4. As a starting point, the modeler would construct the process model graph, entering it graphically using standard schematic capture techniques. As explained previously, each node of the graph represents a process and the arcs between nodes the signal flow between processes. Once the process model graph was complete, the Modeler's Assistant could then lead the modeler through a menu-driven prompt/response process which would define the functions of the processes. Each node has arcs (signals) into it and arcs (signals) out of it. Thus an I/O relationship between the inputs and outputs is implied. A library of standard functions would be available from which the user could choose. They could perform general logic functions expressed as truth tables or finite state machines. In this case the user would provide the table entries and the system would generate the equivalent VHDL. Or the functions could be more specific (e.g., decoding, edge detection, parallel-to-serial conversion, etc). Thus the preliminary work in the design of the assistant would be the definition of a set of primitive functions from which all models could be constructed. Once the modeler selected a primitive function, he or she would be prompted for such things as how the function is connected to the process node inputs and outputs, and model parameters such as delay. Perhaps the model generated would be incomplete (i.e., the modeler would still have to code certain sections manually). Nevertheless, a great deal of effort could be saved by this appraoch and also a discipline in model structure could be enforced.

A Modeler's Assistant is now under development at Virginia Tech, and with the increased popularity of hardware description languages, other efforts are no doubt in progress elsewhere. We look forward to the time when these very useful tools will be available.

OTHER ISSUES

There are other important issues that relate to the use of hardware language models. We discuss three here:

1. *Silicon compilation.* Here the model is used to compile a silicon layout that implements the modeled logic function. Silicon compilation is not new, but in many cases the hardware description language has been tailored to the available VLSI layout macros. A more difficult problem is the compilation of silicon from a general hardware description language such as VHDL. There are a number of major issues here. First is the question of the hardware implied by the language constructs. AHPL is very nice in this regard. Combinational and sequential logic elements are identified in the declaration section and the statement sequence maps directly into a particular control structure. VHDL, on the other hand, was designed to be a general language and implies no particular hardware structure, but this makes it more difficult to compile. Also, except in the case of signals used in guarded signal assignments, VHDL does not identify memory elements. These deficiencies can, however, be corrected by the appropriate use of commenting.

A second issue is modeling style. The situation here is similar to the fault-modeling problem. Data flow models imply more circuit structure and are thus easier to compile than algorithmic models using loops and wait statements.

2. *AC logic test development.* In this book we have emphasized the use of the timing modeling capability in today's hardware description languages. One major application area for this capability is the generation of test patterns for AC logic tests (i.e., logic tests in which time response as well as proper value is monitored). Current logic testers have the capability to impress thousands of test patterns on a logic circuit in a very short period of time and check for proper response. The question is: How are all these test patterns developed? One approach is to use chip-level models to develop the AC response for an existing set of test vectors for which the numerical results are known but time response is not. A second approach would involve defining the concept of a timing fault. Tests then could be generated to detect these faults, thus providing another source of test patterns for AC logic testors.

3. *Model validation.* A problem related to the use of timing modeling is that of model validation. Chip-level models of large devices are complicated and their validation is a nontrivial task; for example, checking out a model of a microprocessor is akin to checking out a microprocessor itself. One subjects the model to a successively more difficult sequence of diagnostic tests, and when it passes all of them, the model is "blessed" as being correct. Unfortunately, this method can only be used to check for correct answers on a DC basis (i.e., no timing can be verified). A concept known as "real chip" simulation has emerged recently whereby a real logic chip is interfaced to a simulator and acts as one of the components in the simulated system. Such an approach could be used to validate the timing of a simulation model by doing an on-line comparison of the time response of a real chip and a corresponding chip-level model of the same device. A question that might be raised is: Why bother to validate a model of a device when the real chip can be used in simulation? The answer is that a simulation model allows for control of the response (e.g., time delays can be varied), whereas with the use of the real chip, one has no control over the internal characteristics of the model. Thus real chip simulation is not suitable for all simulation applications.

SUMMARY

In this chapter we have discussed other issues related to chip-level modeling and the use of hardware description languages. We showed that one can define chip-level fault models and develop test generation techniques for those fault models. A method for measuring the fault coverage of test sets developed at the chip level was presented.

The importance of performing chip-level simulation in parallel was discussed and an architecture to perform this process was presented. A key element of the architecture was the mechanism by which different processors could simulate different parts of the design, yet still be synchronized to the correct simulation time.

The generation of chip-level models is a significant programming task. A Modeler's Assistant could aid in this process.

A major application of hardware description language descriptions is silicon compilation, and some of the characteristics needed by the HDL and model structure to aid this process were discussed. Finally, we discussed the use of the timing modeling capability in hardware description languages in generating tests for AC logic testers. A related problem was the validation of chip-level models.

All the issues discussed in this chapter are important to the development of the chip-level modeling process and the use of hardware description languages.

BIBLIOGRAPHY

Barclay, Daniel S.,"An Automatic Test Generation Method for Chip-Level Circuit Descriptions," Masters thesis, Virginia Polytechnic Institute, January 1987.

Barclay, Daniel S., and James R. Armstrong, "A Heuristic Chip-Level Test Generation Algorithm," *23rd Design Automation Conference*, pp. 257-262, June 1986.

Roth, J. Paul, "Computer Logic, Testing, and Verification," Potomac, Maryland: Computer Science Press, 1980.

Roumeliotis, Manos,"Multiprocessor Logic Simulation at the Chip Level," Doctoral dissertation, March 1986, Virginia Polytechnic Institute.

Roumeliotis, Manos, and James R. Armstrong, "HDL Modeling for Process Oriented Simulation," *8th International Conference, Computer Hardware Description Languages and Their Applications*, the Netherlands, April 1987.

Roumeliotis, Manos and James R. Armstrong, "A Multiprocessor Approach to Functional Parallel Simulation," *International Conference on Parallel Processing*, pp. 462-464, August 1984.

Shen, J.P., W. Maly, and F.J. Ferguson, "Inductive Fault Analysis of MOS Integrated Circuits," *Design and Test of Computers* Vol. 2, No. 6, pp. 13-26, December 1985.

POSTSCRIPT

The purpose of this book has been to define a higher-level view of digital systems. In defining this view we introduced the concept of a chip-level model—a model of a device that reproduces the I/O response of the device without modeling its internal gate structure. A major requirement for these models is that the propagation delay along internal data paths be modeled accurately. We saw that this requirement was met by the use of the process model graph as the basic model structure.

VHDL has been used as the hardware description language for chip-level modeling. In fact, a large percentage of the book has been expended in explaining VHDL. This is inevitable, as any approach to modeling is tightly coupled to the constructs available in the hardware description language. VHDL is particularly well suited to the chip-level modeling application because of its fundamental use of the process. At times, VHDL descriptions may appear verbose. However, this is a necessary result of the fact that the language attempts to achieve a high degree of accuracy in modeling hardware; and to do this without sacrificing generality.

Throughout the book, the modeling process has been illustrated by a good number of examples. First, basic models for combinational and sequential logic circuits were presented. Next, the chip-level modeling process was illustrated for three different circuits of increasing complexity. Finally, the concept of system modeling at the chip level was illustrated by means of a complete system example which included the models of the generic types of logic found in computer systems. It should be evident from these examples that there are many styles of modeling possible, but as the use of chip-level models increases, standards will have to be developed to ensure that models meet certain requirements. For example, one might want to know whether a model has delta delay timing, simple I/O timing, or data path timing. All three types of models have their useful place in the modeling process, but it should be clearly specified to which timing standard the model adheres.

Other aspects of modeling style are important and relate to the intended use of the model. For example, we recommended that data flow behavioral descriptions be used instead of algorithmic descriptions for models intended for fault modeling or silicon compilation. Finally, it should be evident that model features are recurrent, and thus it should be possible to define a structured method of model development, particularly for generic types of logic. In this regard, the concept of the Modeler's Assistant discussed in Chapter 6 could be of particular value.

To be able to model faults at the chip level and develop tests for those faults is a required capability. Thus research in this area is very important and will no doubt continue.

The main purpose of chip-level modeling is to provide system designers with a proper perspective from which to view the operation of digital systems. The advent of hardware description languages has given us a major tool for performing this process. As discussed here, other tools are required. It will be interesting to track their development.

INDEX

A

abstraction hierarchy, 1
AC logic tests, 141
algorithmic body, 17
architectural bodies, 15, 16
arithmetic adding operators, 29
arithmetic types, 24
array type, 26
assertions, 67, 73, 77
asynchronous circuit, 46, 48
attributes, 31, 50, 67, 73

B

behavioral domain, 2
behavioral representation, 4
BIT type, 24
BIT-VECTOR type, 24
block statements, 22
BOOLEAN type, 24
bottom-up design, 7

bundled signals, 85
busing, 88

C

case statement, 35
CHARACTER type, 25
chip-level, 5, 57
chip-level model, 57
circuit level, 4
combinational logic, 43
compiled simulators, 10
complexity, 1
component interaction, 55
composite type, 25
concatenation, 48
concurrency, 38
constants, 29
control fault, 128
counter, 46, 75